Aaron Bank and the Early Days of
U.S. Army Special Forces

by
Darren Sapp

May, 2011
www.darrensapp.com

Cover Design:
Vivid Covers
www.vividcovers.com

Formatting:
Christine Borgford of Type A Formatting
www.typeaformatting.com

ALSO BY DARREN SAPP

Fire on the Flight Deck
The Fisher Boy
Special Force: A World War II Commando Novel

Discover more at *www.darrensapp.com*

Contents

INTRODUCTION

ON A WINDY, rainy day in November 2005, the John F. Kennedy Special Warfare Center and School renamed its academic training building Colonel Aaron Bank Hall. Approximately 200 members and former members of the Special Forces community, as well as civilians, attended the ribbon cutting and enjoyed a thankful speech in English, French, and German by Mrs. Catherine Bank.[1] Although far from a household name, Aaron Bank is famous among the U.S. Army Special Forces, commonly known now as the Green Berets. What merits this honor? Why is he lauded as a Special Forces legacy? How did he earn the title, Father of the Special Forces?

Aaron Bank's exploits rank him among the elite of America's soldiers. Fluency in French and German earned him ideal candidate status as an officer in the newly formed OSS (Office of Strategic Services), the precursor to the Central Intelligence Agency. During World War II, President Franklin D. Roosevelt tasked William Donovan to create one entity to coordinate United States intelligence gathering and conduct clandestine military operations. A crucial element of those operations involved small teams known as Jedburghs that combined one American or British officer, one French officer, and an enlisted wireless telegraph operator to work with the French Resistance. Bank was one of those Jedburghs. This is the story of the early days of U.S. Army special warfare through the life of Aaron Bank from his service in World War II to the official creation of the U.S. Army Special Forces in the early 1950s.

In a sense, this is a biography, but one that focuses on the

1. Joe Healy, "SWCS Dedicates Bank Hall," *Special Warfare* 19, no. 1 (January/February 2006): 6, *http://www.soc.mil/swcs/swmag/Archives/06_JanFeb.pdf* (accessed August 26, 2010).

period of 1944 to 1952 where a few men championed the need for a specialized unit capable of training indigenous forces for irregular warfare, intelligence gathering, and psychological operations. While many deserve their place as a "founding father" of U.S. Army Special Forces, nearly every published work on Special Operations as well as memoirs from other OSS veterans, mentions Bank's contributions. By chronicling the selection of men like Bank, the training they endured, their missions, and their desire to make this type of force permanent in the U.S. Army, the reader will have a clear understanding of the early days of U.S. Army special warfare. Additionally, the compelling contribution of Aaron Bank will serve as a guide through this piece of military history.

What exactly does the term *special warfare* describe? Appendix A includes a glossary of terms and abbreviations but it is important to define these terms as they pertain to a particular section of this thesis. For example, fighting units can contain elements of guerilla warfare, unconventional warfare, partisan warfare, irregular warfare, psychological warfare, and special warfare, but each of those terms can also stand on its own. Examples of names for members of these groups are guerillas, special operators, or commandos. Since conventional warfare is predominately used for that type of warfare, *unconventional warfare* is the best term to describe military or non-military operations that fall outside the conventional use of infantry, armor, air, and naval forces. French Resistance forces in World War II mostly performed unconventional warfare and were composed of civilians, regular French infantry, communist fighters, and even Germans seeking an end to Nazism.

On February 21, 1995, U.S. Representative Ron Packard entered into the official Congressional Record remarks acknowledging the contributions of Aaron Bank, whom he called the Father of the Green Berets.[2] The *Green Berets* most commonly refers to the

2. House, Extension of Remarks, Tribute to Col. Aaron Bank, by Ron Packard, 104th Cong., 1st sess., H. Res. E 389, http://www.gpo.gov/fdsys/pkg/CREC-1995-02-21/pdf/CREC-1995-02-21-extensions.pdf (accessed August 9, 2010).

U.S. Army Special Forces (upper case), very specifically a branch within the U.S. Army that trains and works with indigenous forces in numerous capacities from combat operations to humanitarian endeavors. U.S. Army special forces (lower case), U.S. Army special warfare, and U.S. Army special operations are nearly synonymous and can refer to other units within the army such as Rangers, SOAR (Special Operations Aviation Regiment), Delta Force, etc. While many call Bank the "Father of the Green Berets," most soldiers will prefer the "Father of the Special Forces." Indeed, they might even mention that Major General Robert McClure is the "Father of U.S. Army Special Warfare." Due to the vast body of written material on the subject of these various forces, the reader can infer that standard definitions for these terms do not always exist.

The armed forces of nearly all the developed countries contain elements of unconventional forces although that has not always been the case. Memoirs from unconventional force veterans typically contain stories of friction between them and conventional force leaders. Delta Force operators were extremely frustrated with General Norman Schwarzkopf during Desert Storm over his hesitancy—and some would suggest dislike for special operators—to employ them in pertinent situations.[3] This friction existed in World War II, but it was the first major conflict where unconventional forces were planned, recruited, trained, and employed on a large scale with many successful missions. Chapter one will further describe this type of warfare and how World War II affected it. Specifically, the OSS unit known as the Jedburghs is a shining example of unconventional warfare aiding conventional warfare during the Normandy invasion.

While the exploits of these forces in the field provide entertaining reading, the training they endured is interesting on numerous levels due to the broad capacity of knowledge required of them.

3. Douglas C. Waller, The Commandos: The Inside Story of America's Secret Soldiers (New York: Simon & Schuster, 1994), 273.

For U.S. forces, this was new territory, and the British provided instrumental expertise based on their experience with commando operations. Some of this training benefited pure American units but others integrated carefully chosen American soldiers with British soldiers to form cohesive fighting units. Chapter 2 describes the formation and development of these units while chapter 3 transitions to a discussion of the indigenous forces and environment that formed the basis for Jedburgh operations. Aaron Bank's selection as a Jedburgh took him on a journey from the United States to England, Paris, and Indochina.

Bank's leadership skills and performance in training earned him command of Team Packard, a Jedburgh team that parachuted into France after D-Day with a French officer, French wireless telegraph operator, and a vast amount of supplies and equipment. While nearly 100 Jedburgh teams operated in France, chapter 4 will highlight the accomplishments of Team Packard in harassing and fighting Germans in partnership with the French Resistance. Following Bank's time in Team Packard, a mission to capture Adolf Hitler and other high-ranking Nazi officials developed. Called Iron Cross, the order came directly from William Donovan and consisted of Bank recruiting Wehrmacht soldiers holding anti-Nazi sympathies or simply those seeking a speedy end of the war who would infiltrate Nazi units and aid in kidnapping operations. This mission, outlined in chapter 5, as well as his work in Indochina moved Bank from his position as a Jedburgh to an OSS operator.

These World War II experiences, followed by service in Korea, shaped Bank's view of the need to make an OSS-type force permanent in the United States Army. The final chapter entails the struggle and subsequent victory of key special operations veterans to realize this force. In addition to the previously mentioned McClure, Russell Volckmann and Wendell Fertig—veterans of guerilla warfare in the Philippines—played integral roles in the creation of the U.S. Army

Special Forces. This thesis does not seek to promote Aaron Bank as more important or more responsible for that accomplishment, but rather to show him as one of the main proponents of the need for a permanent unconventional unit.

Bank's memoir, *From OSS to Green Berets: The Birth of Special Forces*, limits the events of his life to military related matters during World War II and a few brief periods later. In addition, he cites no other documents or secondary materials but rather provides a narrative based on memories of those events. Bank, like many other OSS veterans that wrote memoirs, excluded many pieces of information due to their classified nature. However, it should be noted that the information these OSS veterans did share in their memoirs was extremely accurate, a testimony to their training in memorization of intricate details to avoid having to write them down and make them potentially available to the enemy in the event of capture. Fortunately, the National Archives opened previously classified OSS personnel records and other materials in 2008 allowing researchers access to thousands of invaluable documents. In particular, the many Jedburgh team reports reveal an enormous amount of detail regarding their work with the French Resistance, and Bank's military personnel records provide many details of his life before World War II.

The son of parents of Russian heritage, Aaron Bank entered this world on November 23, 1902, in New York City. While many Jedburghs followed the traditional route of formal higher education resulting in a military commission, Bank attended public school until he was fourteen and left with no high school diploma. In 1921 and 1922, he attended a Swedish Gymnastics technical school in New York and completed a course qualifying him as a director for physical training. His father died while Bank was an infant and his mother worked as a language teacher. His immigrant family routinely spoke French and German with him during his childhood.

Throughout his life, physical fitness and language learning would hold much interest for him. During the 1930s, Bank worked in two different occupations, first as a swimming and physical fitness instructor in Biarritz, France and then later in various property management roles in Miami Beach, Florida. In 1938, he spent three months traveling in Switzerland, England, Germany, and Belgium, no doubt honing his language skills.[4] In a 1968 *Los Angeles Times* interview, Bank stated that lifeguarding became a career with winter work in the Bahamas and summer work in France including many trips "in and out of Europe."[5]

Interestingly, Bank's own family acknowledges the ambiguity of his life leading up to 1942. His daughter Linda asked her father about his activities prior to World War II, and his answer was, "Tougher people than you have tried to get that out of me." She, like her mother, suspects that he might have already been involved in some sort of intelligence work.[6] The modern armed forces and intelligence networks typically choose applicants for various abilities and then train them in foreign language but in the era of World War II, it was reversed. Did Bank already have experience in intelligence prior to the war? We may never know. We do know that like many men, albeit many younger than he, Bank volunteered for service in the United States Army, August 19, 1942. Just shy of his fortieth birthday, he described himself as 5'8" tall, 155 lbs, with fair complexion and a sturdy build.[7] Indeed, everyone that describes Bank physically confirms a sturdy build and model of physical fitness well into his nineties. Although not known at the time, Aaron Bank was just the kind of soldier the newly created OSS would be looking for.

There are three primary reasons why this story is not only

4. Records of the OSS. RG 226. Entry 224. Box 34. Personal File: Bank, Aaron. National Archives. College Park, MD.
5. "Saving Lives, Destroying Them— All in His Colorful Past," interview by Gordon Grant, *Los Angeles Times* (Los Angeles), August 11, 1968.
6. Linda Ballantine, telephone interview by author, August 18, 2010.
7. Records of the OSS. RG 226. Entry 224. Box 34. Personal File: Bank, Aaron. National Archives. College Park, MD.

important but also compelling. First, while largely known as a conventional war, World War II saw widespread use of unconventional forces; indeed their organization and implementation as a recognized complement to conventional forces had its birth during this time. This is similar to the birth of armored warfare or any other major military branch. Second, we are entering a period where World War II veterans are dying at extremely high rates, and our opportunity to interview them and those that served with them is dwindling. It is not until we delve into these stories that we know the correct questions to ask and can then enjoy the treasure trove of oral history these questions provide. Third, as someone known as the Father of the Special Forces, a force recognized around the world, Aaron Bank certainly merits having his story told.

Chapter 1

A NEW KIND OF WARFARE

MANY VIEW THE practice of ranks of Revolutionary or Civil War soldiers standing still and exchanging volleys as suicidal. What most do not understand is that there was much more order and reasoning behind this practice such as quick mobility of small squads and concentration of fire. In addition, the elements of chivalry guided commanders to conduct war in a humane and proper manner, whether pitched battle or a surprise attack. Prior to World War II, unconventional forces rarely enjoyed continual recognition as regular military forces in most countries, particularly in the United States. Units sometimes organized for a specific mission, but their status was typically short-lived and the units disbanded after the mission was complete. More commonly, unauthorized groups developed, defending their home region through guerilla activity. The pro-Union Kansas Jayhawkers and pro-Confederate Missouri Bushwackers represent two guerilla elements that often clashed in the Kansas-Missouri border wars that extended into the American Civil War. Union generals frowned upon the Jayhawkers disliking the improper way in which they carried out war. Robert E. Lee never openly supported the Bushwackers but appreciated the tens of thousands of Union troops typically engaged in response to potential Southern guerilla activity.

Aaron Bank described the view of regular army men regarding unconventional war as ungentlemanly. Bank's job entailed "teaching men to kidnap, steal, cheat, or kill by the quickest, most ungentlemanly means possible."[8] Certainly all types of warfare have existed since the beginning of time and continue to this day from stone throwing peasants to the most highly trained, elite commando. However, the "regular army men" that Bank mentions is neither a pejorative nor a complimentary term but rather a descriptor for conventional military forces regardless of the time they existed (e.g. the ancient Roman army or the U.S. Army's 7th Cavalry). The shift in philosophy regarding unconventional forces grew out of a modernization of military forces. Improved communication, weaponry, and intelligence networks, coupled with the size and scale of World War II, facilitated a different type of warfare. The art of war now contained a method that would ultimately save human lives while at the same time more efficiently defeat the enemy. To support this efficiency, the higher command recognized formalized unconventional units, many patterned after the British commando units.

A Brief History of Unconventional War
UNCONVENTIONAL WARFARE IS as old as war itself. From the biblical writers to the writings of Josephus, one can find instances that seem beyond the scope of conventional war. Author Tom Clancy writes, "There is no clean division between conventional and unconventional wars."[9] One of the most famous examples of ancient unconventional war, although likely myth, is the story of the Trojan Horse, where men hid in a large wooden horse to conduct a surprise attack on Troy. More recently, King Philipp's War showed modern forces working with indigenous forces to produce a cohesive fighting force. Roger's Rangers was a special forces unit

8. "Saving Lives, Destroying Them— All in His Colorful Past."
9. Tom Clancy, Carl Stiner, and Tony Koltz, *Shadow Warriors: Inside the Special Forces* (New York: G.P. Putnam's Sons, 2002), 59.

during the French and Indian War conducting reconnaissance operations in which some of the greatest challenges came from the environment. Far from home, these soldiers trained to fight the weather, lack of food, and unknown terrain, as much as they did the enemy. Francis Marion's unit during the Revolutionary War represents a perfect example of guerilla warfare where knowledge of the terrain and use of local civilian intelligence networks made for an effective force multiplier.

While many will classify some of the previously mentioned units as elite, they still operated beyond the conventional role. However, famed unconventional units in the Civil War do not hold elite status. At the onset of the War Between the States, many young men left the regular army and chose guerilla warfare and a life on the run. Much less organized than their Union counterparts, Confederates operated in a much more haphazard manner. They would fight in a small skirmish or conduct an act of pillaging, then escape several miles on horseback. A string of neighbors and relatives would provide places to hide, fresh horses, food, and alibis. If they were injured, a doctor, sympathetic to the cause, gladly gave them care and then sent them back into the field with fresh supplies. Northern forces found it extremely difficult to control guerilla activity due to the deep sympathies of the local population for the guerillas.

William Quantrill is the name most associated with these guerillas, developing a cult following and organizing groups to carry out sabotage against abolitionists forces.[10] However, the names Bloody Bill Anderson, Jesse James, and Cole Younger represent typical families who suffered terribly during the war. The Union forces began a campaign to rein in the guerillas through their families. After confiscating and burning many of the guerilla's farms, they arrested and imprisoned many of the female members of their

---

10. Richard S. Brownlee, *Grey Ghosts of the Confederacy* (Baton Rouge: Louisiana State University Press, 1986), 57.

families on the charge of aiding the guerillas. Both sides agreed that raids conducted as guerillas violated the rules of war and non-combatants suffered terribly. While tactically, the Southern guerillas effectively terrorized many Union towns, they were no match for the conventional Union forces that ultimately extinguished them.

Despite the efforts of guerillas, infantry continued to bear the heaviest casualties in battle well into the next century including the horror of trench warfare in World War I. Over the next twenty years, there was a huge technological shift in military arsenals on the sea, air, and land. Infantry always has and likely always will exist, but with the advent of World War II, first strike capability considered the use of bombing before infantry. Tanks paved the way for foot soldiers, battleships softened defenses, and bombers destroyed war-making factories from the air. Enough influential leaders and commanders embraced the idea of organized unconventional forces as viable assets for efficiency of force and the saving of human lives that intelligence networks, special training, and billets opened to form organized units.

In a letter to British SOE (Special Operations Executive) commander Collin Gubbins regarding the work of Jedburghs, Dwight D. Eisenhower praised them. He writes,

In no previous war, and in no other theatre during this war, have resistance forces been so closely harnessed to the main military effort. While no final assessment of the operational value of resistance action has yet been completed, I consider that the disruption of enemy rail communications, the harassing of German road moves and the continual and increasing strain placed on the German war economy and internal security services throughout occupied Europe by the organized forces of resistance, played a very considerable part in our complete and final victory.[11]

He goes on to attribute the planning, training, and

---

11. Dwight D. Eisenhower to Collin Gubbins, May 31, 1945, *http://www.spartacus.schoolnet. co.uk/SOEgubbins.htm* (accessed August 7, 2010).

implementation of an organized military force to work with citizens under occupation as the means by which the resistance forces organized. Although the United States and Britain led the way in the organization of unconventional forces, the Wehrmacht first implemented them, designed for their special need in World War II, support of the Blitzkrieg.

Special Needs of World War II

THE OFFENSIVE ONSLAUGHT known as the Blitzkrieg effectively overwhelmed everything in its path through air power and armor followed by infantry. Preceding the invasion force were German commandos specially trained to control bridges and other vital areas in advance of the main force. Due to their success, these commando units became a permanent force in the early days of World War II.[12] Prior to military action in Africa, no other special forces earned any permanent recognition in any other armies. British commandos, along with U.S. Army Rangers, sustained horrible losses in the raid on Dieppe, casting further doubt on special forces. Churchill, however, stood by commando operations as a necessary and viable option. Some historians discount the overall achievements of special forces units in World War II but that is like suggesting any other ancillary unit to conventional forces made no significant contribution. When one looks at the many facets within the repertoire of unconventional forces, it becomes clear that their contribution was vital to the defeat of the Axis powers. Special needs such as intelligence, deception, tactical missions, and collaboration with partisans during World War II influenced commanders to create permanent solutions rather than temporary ones.

At the forefront of every military unit is intelligence, whether land, sea, or air forces. Typically, units such as a battalion will assign an officer to fill this billet, but at the onset of World War II, Great

---

12. David Thomas, "The Importance of Commando Operation in Modern Warfare 1939-82," *Journal of Contemporary History* 18, no. 4 (October 1983), 691.

Britain and the United States created the SOE and OSS, respectively, as special units to oversee all intelligence and covert operations. Intelligence became an enormous task with ombudsmen acting as intermediaries between units. In addition, the art of deception often finds itself coupled with intelligence work. The numerous personal messages broadcast by the BBC to the French Resistance relayed valuable messages to the Allies while at the same time confusing the Germans, in particular during the days leading up to and just following D-Day.[13] OSS Jedburgh teams, tasked with training the French Resistance, harnessed intelligence strategically working like the aforementioned ombudsmen and relying on those BBC messages to implement their forces.

Reconnaissance means a few different things, but essentially its function offers commanders the information they need to plan covert or offensive operations. Scouts or reconnaissance parties advance ahead of the main force and return or relay the intelligence information, which might include enemy strength, terrain obstacles, civilian concerns, etc. The sheer geographical vastness of World War II presented a unique problem where the Allied forces operated in numerous unfamiliar areas. In 1943, the 29th Ranger Battalion conducted harbor reconnaissance on the Norwegian coast with few tactical results but this exercise provided valuable experience.[14] Invasions in North Africa, the Pacific, and Normandy all benefited from the specialized reconnaissance performed by these special Ranger units.

Today's Ranger units, Marine Force Recon, or Navy SEALs are tasked with small-unit tactical operations much like the work done by several airborne and commando units during World War II. Due to the exceptional plan of deception, the Germans held back much of their infantry and armor not knowing if the Allied

---

13. Jock Haswell, *The Intelligence and Deception of the D-Day Landings* (London: Batsford, 1979), 172.
14. David W. Hogan, *U.S. Army Special Operations in World War II* (Washington, D.C.: Center of Military History, Dept. of the Army, 1992), 39.

invasion of Europe would occur in Normandy, Brittany, or even the Pas de Calais. Once the Normandy invasion commenced, the Allied forces assumed the Germans would commit all available Panzer divisions to the Normandy front. Thus, the control of bridges would prove vital to success or failure. Just minutes after midnight, British Airborne commandos, led by Major John Howard, landed at Pegasus Bridge via glider as the first unit on the ground in Normandy on D-Day. Historian Stephen Ambrose described their heroic capture and holding of the bridge as one of the single greatest small-unit actions of the war, and indeed, their failure possibly might have meant disaster for the entire invasion.[15] The tasks of preparing invasions routes, softening defenses, and operating far ahead of conventional forces lie at the heart of most special forces and greatly shaped Aaron Bank's views during the formation of the United States Special Forces.

As the Germans invaded and occupied several European nations, their confident and jovial nature as conquerors proved horribly unsettling to Frenchmen as evidenced by one of the most famous film clips of World War II of a French man crying as the Nazi's marched into Paris. The advantages held by the Wehrmacht in France lay in the close proximity to Germany allowing for control of railway and supply lines, as well as the overall familiarity of French language, culture, and geography to many Germans. However, they failed to realize the French resolve and the Allied collaboration with the French Resistance. Although the Germans would violently retaliate, Jedburghs and OSS Operational Groups conducted hundreds of acts of harassment such as disruption of railway lines, deception of military strengths, and sharing of false intelligence. Where pockets of resistance existed, Allied forces attempted to employ special force teams to aid in harassment of the enemy.

15. Stephen E. Ambrose, *Pegasus Bridge: June 6, 1944* (New York: Simon and Schuster, 1985), 182-83.

Throughout the world, occupied peoples acted either in heroism, such as the Filipino resistance against the Japanese, or capitulation to the enemy, such as the Vichy French embracing the Germans. Determining whom to believe and trust proved a great difficulty. The town baker might be a valuable courier of information for the French Resistance or a tool of misinformation acting for the Germans. OSS Jedburgh Teams proceeded with extreme caution working with the French Resistance. Their missions were dangerous from the standpoint of parachuting behind enemy lines and combat, but the art of working with and relying on indigenous forces could be deadly. Jedburghs trained intensely for this work and enjoyed moderate success and occasional frustration. The U.S. Army would spend 8 years following D-Day before permanently establishing the special force team to do this work and naming Aaron Bank its first commander.[16]

Although the "special needs" of World War II have many different nuances, one final one is worthy of discussion. The rescue of prisoners of war proved a difficult task in Germany but a plan to rescue the Allied soldiers held at Cabanatuan, Philippines, in 1945 succeeded. The elite U.S. Army 6th Ranger Battalion succeeded due to collaboration with highly motivated Filipino guerillas whose knowledge of the terrain and jungle combat skills provided invaluable support.[17] The many facets of unconventional war found a permanent home in World War II as war planners acknowledged their need in future wars. Psychological, reconnaissance, and tactical operations remained too specialized a task for conventional units to spend their time, training, and man power while dedicated units could devote all their training on those special tasks.

---

16. Aaron Bank to Mrs. Beverly E. Lindsay, February 27, 1973.
17. Hampton Sides, *Ghost Soldiers: The Forgotten Epic Story of World War II's Most Dramatic Mission* (New York: Doubleday, 2001), 75.

From a Special Office to a Special Force

WERE THERE AN official title for guerilla war and commando expert at the onset of World War II, British officer Sir Colin Gubbins would have held the honor. Veteran of World War I and the Russian Civil War, he had written two short manuals on guerilla warfare and leading partisans.[18] As head of the SOE, Gubbins and other British leaders created the integral intelligence and covert network that shaped and influenced other Allied organizations, in particular the United States. Two organizations, MI5 (Military Intelligence-Section 5) for domestic intelligence and MI6 (Military Intelligence—Section 6) for foreign intelligence, predated the First World War but Sir Winston Churchill recommended the United Kingdom create one office overseeing military intelligence and covert operations. Thus, with the birth of the SOE, the British committed themselves to military covert operations and the United States would soon follow.

William Donovan wore many labels: millionaire, attorney, friend of President Roosevelt, and Congressional Medal of Honor winner for leading an attack during World War I. Donovan's views on the emerging war in Europe, coupled with his "organizational vision," impressed the President, and he delegated him control of the newly created COI (Office of the Coordinator of Information).[19] Donovan took full control of the office as a means to collect and analyze data and perform other supplemental activities, very similar to the operations he previously observed in Europe. Donovan felt the office would be crucial in preparing the European Theatre for a potential Allied invasion and foresaw the likelihood that it would grow far beyond its initial conception.[20] Not everyone embraced

18. S.J. Lewis, "Jedburgh Team Operations in Support of the 12th Army Group, August 1944," CGSC—Command and General Staff College, *http://www.cgsc.edu/carl/resources/csi/Lewis/Lewis.asp* (accessed August 17, 2010).
19. R. Harris Smith, *OSS: The Secret History of America's First Central Intelligence Agency* (Berkeley: University of California Press, 1972), 1.
20. David W. Hogan, *U.S. Army Special Operations in World War II*, 7.

Donovan or the new office due to varying viewpoints on a civilian office conducting military operations.

The COI was converted to the OSS in June 1942, under the authority of the Joints Chiefs of Staff and Donovan immediately sought to expand the nature and mission of the new office. In addition to intelligence analysis and gathering, new departments formed, such as research and development for weapons and equipment, counterespionage, psychological propaganda, and maritime units.[21] The other major departments consisted of special operations that included small unit military tactics, espionage, raids, guerilla warfare, reconnaissance, etc. The vision of Donovan for the specific organization of this office was fairly clear, but the mission was absolutely clear—conduct irregular warfare to harass and demoralize the enemy while supporting the Allied conventional forces. Finding the right people to perform these tasks became Donovan's next major goal.

OSS operators might work solo or in teams. They could be male or female. They would likely coordinate with soldiers or civilians of many nations. An often-attributed quotation from Donovan was that he wanted "an Ivy League Ph.D. who could win a bar fight."[22] While that quotation is very descriptive, an interpretation might be that the OSS wanted people from all walks of life, but they needed to excel in education, daring, and foreign language skills. They should be teachable, mature, cultured, and cosmopolitan. OSS recruiters scoured college campuses and military bases seeking civilian and military recruits. Twenty-year-old college student Barbara Hans met with a recruiter at Smith College who asked her questions such as, "Do you like puzzles?" After acceptance, she trained in Washington, D.C. as a communication specialist and served in Ceylon in a quality control capacity, ensuring American

21. Patrick K. O'Donnell, *Operatives, Spies, and Saboteurs: The Unknown Story of the Men and Women of World War II's OSS* (New York: Free Press, 2004), xv-xvi.
22. Samuel A. Southworth, *U.S. Special Warfare: The Elite Combat Skills of America's Modern Armed Forces* (Cambridge, MA: Da Capo Press, 2004), 132.

codes met certain standards. Although a civilian, she experienced the great adventure of military life "without a .45 pistol."[23]

Aaron Bank responded to a recruiting announcement seeking soldiers with foreign language skills, while stationed at Camp Polk, Louisiana. Intrigued at the idea of overseas service, he eagerly accepted the offer to join the OSS. His fluency in the requisite French was enough, but his competency in German was a huge bonus. Recruits like Bank served in numerous special operations units throughout both theatres, but there were two major groups that operated in Europe in the early days of the OSS. The OGs (Operational Groups) performed military tactical operations much like Army Rangers but with a size of force of around 30 soldiers, and the Jedburgh Teams consisted of three men that organized, trained, and led units of the French Resistance for covert operations.

Jedburgh Team Alexander's mission order was to act as a liaison between several Allied groups operating in central France, set-up and maintain communication links with London, and "in particular they were to harass enemy movements on roads and railways."[24] Bank's Team Packard mission statement is very similar and typical to other Jedburgh teams. While the art and skill of persuasion, camaraderie, culture, and language played key roles, practical elements of communications, military tactics, and weapons needed mastering as well. In just one year, the OSS had morphed from a small intelligence office to a major covert operations command with thousands of personnel. The time to develop every element of this organization should have taken years, but fortunately, the OSS could look to the British to help train these new recruits as they had Darby's Rangers the previous year in Scotland.

23. Barbara Hans Waller, telephone interview by author, April 18, 2010.
24. "Report of Jedburgh Team Alexander," digital image, Operation Jedburgh, *http://www. operationjedburgh.com* (accessed August 9, 2010).

Chapter 2

COMMANDOS IN TRAINING

TO TRAIN A soldier in covert operations such as espionage, hand-to-hand combat, and various weapons is one level of combat training. Training those same soldiers how to teach these skills to partisans is another level. While modern skills training has a depth of experienced instructors, the World War II era depended on people with specific tradecraft such as use of the Sten gun, combat with a dagger, W / T (wireless telegraph) operation, hiding documents, etc. In other words, instructors with no combat or military experience trained Jedburghs on their particular area of expertise. Of the hundreds recruited for training, instructors selected three men per team for approximately 100 teams. Lists of Jedburgh teams will differ due to restructuring, internal team changes, teams renamed, and teams redeployed. No team, however, entered the field untrained in mind or body for the task of leading partisans.

Forty-one year old Aaron Bank was likely the oldest recruit, but also one of the fittest. As a physical fitness instructor, he embraced the training believing it should be rigorous and challenging. Academically, he did not have the Ivy League credentials of the other recruits, but his proven military record of leadership qualified him for a place among these peers.[25] The personalities of these recruits,

25. Records of the OSS. RG 226. Entry 224. Box 34. Personal File: Bank, Aaron. National Archives. College Park, MD.

their willingness to volunteer knowing that many would not sur-
vive, and their motivations for volunteering offer keen insight into
soldiers that served in the most daring of missions in World War
II. Training began in the United States and eventually moved to
England where the Jedburghs learned more in a few months than
they likely ever had in their lives.

Scholarly Fighters

SELECTION TO BEGIN training did not guarantee recruits a spot
on a Jedburgh team, but rather temporary assignment to the OSS.
Those that failed final selection returned to their previous units
in most cases, but some found assignment to operational groups.
Those selected held similar traits. The average recruit had a college
degree, spoke French fluently, held an army rank of lieutenant
or captain, and distinguished himself in some other way such as
athletics. Selection for American recruits was more stringent than
their British and French counterparts because the majority of the
latter had earned their selection through combat experience. The
final step for most American recruits before moving from stateside
service to Europe involved a face-to-face interview with William
Donovan. The imposing Donovan met with Lt. Jack Singlaub com-
mending his training record and assuring the young lieutenant that
combat would be much rougher than training. Perhaps somewhat
naive to the realities or war, Singlaub replied, "Yes sir."[26] Singlaub
would later earn command of Jedburgh Team James.

Recruiters touted the Jedburghs as an elite unit worthy of only
the finest soldiers. They did not mention the expected casualty
rates of 50 to 75 percent or that if captured they would likely be
executed.[27] It is likely that would not have mattered. As Jedburgh
training continued and they became more aware of their mission
and the danger, the records rarely indicate recruits quitting on their

26. Colin Beavan, *Operation Jedburgh: D-Day and America's First Shadow War* (New York:
Viking, 2006), 64-65.
27. Ibid, 49.

own. The pursuit of adventure, the desire to be among the elite, and the necessity to defeat the Axis drove these men. This certainly described Bank, but for others it was more personal. Team Gavin member William Dreux spent his childhood in Paris taking walks through the Luxembourg Gardens with his mother. Images of German infantry goose-stepping through Paris and Nazi officers "strutting arrogantly" through the Luxembourg Gardens disturbed him. Dreux said, "Humiliation sat on Paris like a huge toad."[28]

Dreux spoke French from childhood while others, such as Bank, picked it up later their life. A few excelled in the language during college. All of them had to know French well enough to communicate with members of the French Resistance but many needed to master the dialect to match their *nom de guerre* and cover story. More than simply conversing with their French Resistance fellow soldiers, Jedburghs needed to teach them in elements of war and lead them in combat. Most of these men excelled in education and embraced the priority of training to learn to effectively communicate as well as lead teams. Most of the Jedburgh teams would arrive by parachute far behind enemy lines, sometimes in uniform, highly trained, and set up a command post to lead men they had never met. Jedburgh leaders assumed the boost in morale for the French Resistance would sway them to comply, and most did, but for many teams internal conflicts among Resistance leaders presented problems.[29] The OSS needed men of achievement, with the ability to adapt to ever changing situations.

Every Jedburgh team included a W/T operator, typically filled by an enlisted soldier. While they needed to be in top physical condition, their qualifications were somewhat less demanding requiring a working knowledge of French and the ability to operate the W/T at 15 words per minute.[30] While they spent a heavy portion of their

28. William B. Dreux, *No Bridges Blown* (Notre Dame, IN: University of Notre Dame Press, 1971), 10.
29. Roger Ford, *Steel from the Sky: The Jedburgh Raiders, France 1944* (London: Weidenfeld & Nicolson, 2004), 12.
30. Lewis, "Jedburgh Team Operations in Support of the 12th Army Group, August 1944," 8.

training time on W/T operation, they still participated in nearly all the other required Jedburgh training. Team commanders considered them to be as valuable members of the team as any other. Every team member needed some proficiency at W/T operation should the W/T operated be killed, injured, or captured. To be a Jedburgh demanded courage and versatility.

Camp "B"

THE OSS HAD canvassed military bases, universities, and businesses to find the best and the brightest. Initial training commenced in Maryland at the Congressional Country Club that had recently converted into a temporary military installation like so many other peacetime locations. Candidates participated in a battery of psychological and physical assessments and trained from dawn until long after sunset.[31] Within a month, the recruits moved to a location called Camp "B," sharing space with the newly created presidential retreat. This area would later become Camp David. Although they were from eclectic backgrounds, the men seemed to work well together and thrived in the challenging training environment. Jack Singlaub describes a varying regimen that involved firing a wide variety of weapons, planting explosive charges, navigating a sophisticated obstacle course, and learning clandestine radio procedures.[32]

There is a very real difference between military basic training and elite unit training. Basic training is an introduction to the military designed to transition the recruit from a civilian mindset to a military bearing. That bearing directly aims to ensure a soldier, sailor, or airman can follow orders and perform duties regardless of distractions or personal conflicts. Elite training attempts to simulate combat stress. Combat veterans overwhelmingly agree

31. Will Irwin, *The Jedburghs: The Secret History of the Allied Special Forces, France 1944* (New York: Public Affairs, 2005), 44-45.
32. John K. Singlaub, *Hazardous Duty: An American Soldier in the Twentieth Century.* (New York: Summit Books, 1991), 32.

that simulating combat stress is impossible but putting recruits
through high-pressure situations that exact a mental and physical
toll are very effective in determining who can and cannot endure
the stress of combat. Anyone who has ever read or seen U.S. Army
Ranger training or U.S. Navy Seal Bud/s knows that pushing recruits to nearly unbearable physical exercise results in high dropout
rates. That is exactly what the instructors desire. This concept has
developed over time but during World War II, elite training relied
much more on psychological assessments to determine how men
would handle combat.

For the American Jedburghs, Camp B represented their introduction to elite training. Traditional psychological tests determined
a recruit's tolerance for stress. The desired operator needed to be
a "secure, capable, intelligent, and creative person who could deal
effectively with uncertainty and considerable stress."[33] Although the
men engaged in mock combat operations, the consensus among
the recruits was that the training was not realistic and not preparing
them for the guerilla warfare they would soon experience.[34] No
Jedburghs would report that their time at Camp B was wasted, but
they were eager to move on to the next level. They had proven
they deserved to be Jedburghs and were mentally and physically
prepared for advanced training.

Of all the instructors, nearly every Jedburgh veteran mentions
British Major William Fairbairn who trained them in the United
States and Great Britain. Veteran of the Shanghai Police Force,
Fairbairn's expertise in hand-to-hand combat was unparalleled.
Although he carried an unimposing stature, he regularly threw
around men much larger than he and demonstrated numerous
ways to kill with his bare hands. Fairbairn's methods all considered
close quarters combat as the way in which they would be employed
whether using bare hands, knifes, clubs, guns, or anything else one

33. *Joint Special Operations University and Office of Strategic Services (OSS) Society Symposium: Irregular Warfare and the OSS Model*, Report, 19 (Hurlburt Field, FL: JSOU Press, 2010).
34. Ford, *Steel from the Sky*, 13

could use to defend himself or neutralize the enemy. Fairbairn's 1942 training manual called *All in Fighting* was renamed *Get Tough* for the OSS in 1943. Techniques such as using a chair like a lion tamer, boxing an opponent's ear, and striking with a matchbox in your hands are just a few of the methods that were taught.[35] Along with a fellow instructor, he developed the Fairbairn-Sykes Fighting Knife, a dagger specifically designed for covert operators. In addition to hand-to-hand combat, he taught them instinctive firing. Rather than aiming through sights, the shooter points the weapon toward the target and fires two rounds in an effort to reduce the second or seconds wasted on aiming. Instinctive firing remains the preferred method of elite units today. Jedburghs commonly reference his techniques as groundbreaking and as training that was of great benefit to them. The Jedburghs' training was as advanced as possible for 1943 and they had very capable instructors for their crash course in commando operations.

Ever the physical fitness instructor, Banks expressed satisfaction at the physical stamina the men had achieved but disappointment at the specialized infantry training. Banks had volunteered for guerilla warfare and sought to move beyond standard practices.[36] Since few manuals, no established history of training, and few instructors with combat skills were readily available, the OSS used Camp B to reach a final group of candidates that would likely qualify for Jedburgh teams. The "graduation" from Camp B meant a trip across the Atlantic for further training with the more experienced British Commandos. Unlike Bank, the trip to London was the first overseas trip for many, and some would never return home.

35. Stephen Bull, *Special Ops, 1939-1945: A Manual of Covert Warfare and Training* (Minneapolis, MN: Zenith Press, 2009), 66-67.
36. Aaron Bank, *From OSS to Green Berets: The Birth of Special Forces* (New York: Pocket Books, 1987), 5-6.

Milton Hall

THE JEDBURGHS TRAINED and lived in numerous locations, although Camp B and Milton Hall in England were the two most permanent homes. After a five-day trip across the Atlantic, the men endured train trips and temporary stays before finally arriving at a large property in the country called Milton Hall, just north of London.[37] The sprawling estate, run by the SOE, contained offices, classrooms, and other training facilities. In addition to the American Jedburghs, British and French soldiers who had completed their initial training arrived and began to acquaint themselves with their new surroundings and one another. The Jedburghs were a decidedly British-run operation but one in which the Americans offered complete cooperation. The Americans found the accommodations adequate and were amused that the British apologized for the shortage of batmen, military orderlies serving the officers.

Training continued in much the same way as Camp B with a clear difference in tempo. Mixed in with the American, British, and French officers were enlisted men and Dutch and Belgian soldiers preparing for operations in their countries. The Americans held a great deal of respect for the British officers, acknowledging their combat experience. The fact that they were all one unit, Jedburghs, enhanced the camaraderie. In his memoir, Bill Dreux states that, "We had a bond, a joint purpose, and while this may not have been a mystique, or esprit de corps, since we're not going into combat as a unit, still it was something very real."[38] An even stronger bond developed as they formed the three-man teams. Working behind enemy lines required enormous training and the ability to perform a teammate's task should that man fall. In addition, trust among team members was vital.

The Jedburghs largely chose their own teams in an effort to

37. Irwin, *The Jedburghs*, 59-60.
38. Dreux, *No Bridges Blown*, 50.

form groups that were compatible. The one requirement for each team to include a Frenchman caused some heavy courting. Aaron Bank spent several weekends with French Lt. Henri Denis, treating him to the finest hotels and restaurants.[39] Apparently, many of the teams formed this way, with American and British officers jockeying for their favored French officer. Typically, the two officers would then find a W/T operator that effectively communicated with them. Organizationally, the chain of command remained in a state of flux. British officers commanded Milton Hall, initially Lieutenant Colonel Frank Spooner, and then Lieutenant Colonel G. Richard Musgrave. Training normalized once command of the Jedburghs fell under a special section in London of the SFHQ (Special Force Headquarters) under SHAEF (Supreme Headquarters Allied Expeditionary Force).[40]

The Jedburghs jumping into France carried weapons, rations, first aid kits, and various other supplies. Their priority was of course to get down safely. Many of the men already had jump training but for the Americans it was from the side of airplanes. They would be jumping through the bomb bay doors of converted British bombers at very low altitudes due to the clandestine nature of their arrival in France. Even for experienced jumpers, this required a new skill so the men practiced first from hot air balloons, then in daytime jumps, and finally in nighttime jumps. For Team Frederick W/T operator, Technical Sergeant Robert Kehoe, his initial jump was his first experience in an airplane. "I had the same reaction as most trainees—fear mixed with a sense of terrific excitement."[41] Some men—never able to force themselves into the plane—withdrew from the program, but most persevered and advanced in the training.

Most Jedburghs achieved firearm competency, but the role as

39. Bank, *From OSS to Green Berets*,19.
40. Irwin, *The Jedburghs*, 70-71.
41. Robert R. Kehoe, "1944: An Allied Team with the French Resistance," *Studies of Intelligence*, Winter 98-99, *https://www.cia.gov/library/center-for-the-study-of-intelligence/csi-publications/csi-studies/studies/winter98_99/art03.html*, (accessed September 7, 2010).

teacher and operator meant they needed a wide proficiency with weapons used by various nations and the ability to build and implement explosives. Basic point and shoot firing is one skill, but the adaptability to use a captured weapon requires immense training due to multifarious clip feeds, ammunition, safeties, inherent problems, effectiveness, etc. The reason the AK-47 has been in wide use in the last forty years—particularly for soldiers of limited training, such as guerillas—is its simple application. It is easy to learn to use and clean, and rarely it fails. During World War II, the Sten gun served this purpose and became the main weapon for resistance forces.[42] It was a handheld submachine gun weighing about eight pounds loaded. Jedburghs received much training on the Sten, and it is the weapon most commonly mentioned in their memoirs.

A vital part of the training once teams formed was two-to-three day exercises known as schemes. The team might parachute many miles away, and while completing certain tasks, make their way back to Milton Hall undetected.[43] Although the risk was minor, this was a valuable tool to build team unity, enhance their communication skills, and gain confidence in adaptability. To compound the exercise, the instructors would notify the local home guard to attempt the Jedburghs capture. Some of the teams, such as Bank's Team Packard, took advantage of patriotic civilian's hospitality and sought refuge with a hot meal and warm bed. The next morning after hot tea, seeing their freshly laundered uniforms, the team had to dirty them to fool the umpires of the exercise. [44] The missing element of those schemes was the ability to locate resistance members, organize them, and lead them on mock sabotage operations. That very idea would one day become the Robin Sage exercise among the Green Berets, a crucible-type operation that every candidate must complete.

In addition to all the specific training was the understanding

42. Bull, *Special Ops, 1939-1945*, 146.
43. Irwin, *The Jedburghs*, 65.
44. Bank, *From OSS to Green Berets*, 21-22.

that black operations might be required, such as breaking local laws, deceptive propaganda, forgery, etc. The Jedburghs had to be very flexible in executing their specific missions. With D-Day imminent, each team began hearing of their mission, with most preparing to jump in after the initial Normandy landings. The environment they would jump into was one of chaos, working with people experiencing their fourth year of occupation. Many of those they would command, or at least fight with, had much more combat experience than they had. Appreciating their accomplishments, the Jedburghs needed to hone skills of the French Resistance for a more effective harassment of the Germans.

Chapter 3

INDIGENOUS FORCES

A COMMON MISCONCEPTION among some Americans is
that the United States alone liberated France. Many others will
acknowledge the broader and correct view that the Allies liberated
France, more specifically Europe, from Nazi tyranny but still hold
the misconception that the French did nothing. In truth, hundreds
of thousands of French citizens risked their lives in defense of their
homeland and in aiding the Allied invasions. While many chose
collaboration with the Germans and some simply chose to live life
as best they could under the Nazi boot, a few rose up in rebellion
against the occupiers. Just as every segment of French society
experienced the occupation, people from each of those segments
resisted in varying ways at great peril. Author David Schoenbrun
describes them as "ordinary people who did extraordinary things."[45]

Before the inception of the Jedburghs, the United States never seriously considered harnessing the power of a guerilla force.
Leading up to D-Day, Eisenhower considered activities and coordination with the French Resistance a "bonus," but in a March 1944
memorandum, he stated that, "We are going to need very badly
the support of the Resistance groups in France."[46] This concession
by the commander of all Allied forces in Europe, suggests that

45. David Schoenbrun, *Soldiers of the Night: The Story of the French Resistance* (New York:
Dutton, 1980), 8.
46 Hogan, *U.S. Army Special Operations*, 48.

he saw guerilla warfare as not only necessary, but also a valuable element of the Allied arsenal. By 1944, France eagerly awaited the invasion of Allied forces. Opposing government allegiances, abusive German occupation, and differing political ideologies contributed to a French people in disarray.

Nothing is Lost for France
AFTER INVADING POLAND, September1, 1939, Germany proceeded to take control of the country in a little over a month. Seven months later, France experienced a similar defeat as Germany rolled through Belgium, bypassing the Maginot Line, a defensive barrier built along the German-French border in response to World War I and the potential of a future invasion. France, woefully ill prepared for war, was no match for the modern German blitzkrieg. German soldiers marched through Paris as a sign of extreme humiliation for the French. Adding to the indignity, Hitler forced the French to sign terms of surrender on the very railway carriage at Compiegne, where Germany signed the armistice of World War I. The future of France was in peril with influential political and military members jockeying for position under occupation.

Appeasing the Germans with a puppet regime, Marshal Philippe Petain led the creation of a new government based at Vichy controlling the zone in the south not occupied by German forces. The occupied zone, directly controlled by the Germans, consisted of the northern part of France. In reality, Germany controlled both zones and subjected the French to aggressive new laws, abusive economic policies, forced labor, and persecution of French Jews. The Vichy government created the *Milice*, a paramilitary unit responsible for security and counterintelligence for the Germans against the Resistance. Operating among fellow citizens, their actions produced hatred and fear.[47] While numerous splinter groups existed, French citizens largely associated with one of the

47. Kehoe, "1944: An Allied Team with the French Resistance."

two groups, the Vichy French or the Free French.

Veteran of World War I and the Battle of France, General Charles de Gaulle refused to surrender to Nazi aggression and became leader of the Free French. Escaping to England, he delivered an impassioned speech over the BBC encouraging citizens that "nothing is lost for France" and "France is not alone." Citing the support of the British Empire and the industrial power of the United States, he called on French officers and soldiers to contact him in London. Last, he stated, "The flame of French resistance must not be extinguished and will not be extinguished."[48] Many of the leaders of both the Vichy and Free French provisional governments were heroes of World War I as well as being current military leaders, and this presented a dilemma of loyalty for many soldiers. To join the Vichy French meant freedom from German persecution while joining the Free French might mean death.

Small pockets of partisans sprang up throughout Europe, taking refuge in barns, secret rooms, mountains, and woodlands. Many more languished in prisons as the Nazi jailed thousands when they occupied various countries. Sadly, many Jews, regardless of citizenship, faced deportation to concentration camps. British Prime Minister Sir Winston Churchill urged a de Gualle representative that "the French should now emulate the Yugoslavs and concentrate on setting ablaze the mountain country of South Eastern France from Nice to the Swiss Frontier." He went on to suggest that the Allies could supply the guerillas in France through parachute drops.[49] To "set Europe ablaze" was an expression by Churchill charging the SOE with the mission to harass the Germans through unconventional warfare, and the idea became a rallying cry for many partisan groups, in particular, the French Resistance.

As the German war machine attempted to force young men

48. "Transcipt of Charles De Gaulle 6-18-40 Speech," The Lehrman Institute Public Policy Programs Lehrman Institute Research, *http://lehrmaninstitute.org/history/index.html*, (accessed October 13, 2010).
49. Arthur L. Funk, "Churchill, Eisenhower, and the French Resistance," *Military Affairs* 45, no. 1 (February 1981): 29.

to work in German factories, many fled in small bands throughout France to hide, resist, and fight. Poorly supplied, disorganized, and dependent on local support, they became resistance fighters known as the *Maquis*.[50] Living very much like the Confederate guerillas that served with William Quantrill in the American Civil War, the Maquis began to symbolize the French Resistance. They would eventually become the eyes, ears, and right arm of Jedburgh teams.

Résistance

THE U.S. ARMY manual for guerilla warfare states that "Resistance is the cornerstone of guerilla warfare."[51] Characteristics commonly applied to guerilla warfare might include adherence to political ideologies, domicile protection, survival, and patriotism. While many times isolated from sources of supplies and moral support, guerillas must practice extreme resourcefulness. Guerillas typically operate while surrounded by enemy forces but rarely have a tactical location to protect and use sabotage as their main offensive action, giving them maximum mobility. The French Resistance held all these characteristics. The Jedburghs quickly learned that the French Resistance guerillas had the ability to move secretly and quickly, amidst and around enemy conventional forces. To achieve maximum effectiveness, the French Resistance sought to organize all the small bands that existed throughout France.

While the numerous resistance groups throughout France had a common foe, the *Boche* (a common derogatory term for Germans), varying factors kept them separated, such as trust, secrecy, disagreements on leadership, and differing political ideologies. Two of the largest groups were the FTP (*Francs-Tireurs et Partisans*)—mostly communists whose name implied free or irregular riflemen, and the FFI (*Forces Françaises de l'Intérieur*)—the group consisting of many former soldiers. From his headquarters in London, de Gaulle created

50. Irwin, *The Jedburghs*, 39.
51. United States., Dept. of the Army., *FM 31-21 Guerilla Warfare and Special Forces Operations, 1961*. (Washington: U.S. Government Printing Office, 1961), 5.

the BCRA (*Bureau Central de Renseignements et d'Action*) directing all resistance efforts between the various groups, as well as organizing supplies and missions.[52] While groups typically operated in very low numbers to maintain concealment, one Jedburgh operation had 600 men of various Resistance groups working in unison to harass the Germans.[53] Groups eventually formed into battalions and, on occasion, participated in pitched battle against the Germans.

Resistance fighters, those that supported them, and their leaders faced severe consequences if captured. The Germans did not apply Geneva Convention protections to members of the Resistance and subjected them to unspeakable torture, imprisonment, and execution. Stories abound of faithful Resistance members that endured endless torture by the Gestapo rather than divulging information that might compromise fellow Resistance members. The Germans knew the French Resistance was a valuable intelligence resource for the Allies and suspected all citizens as possible intelligence gatherers. While members of the FTP and FFI participated in "active" resistance, many others participated in passive resistance, such as three elderly sisters who allowed the use of a field near their chateau as a marshalling area for parachute drops.[54] A Resistance force near the small village of Oradour attacked the Germans in a typical harassment operation. In retaliation—and as a warning to further Resistance activities—the Germans massacred 642 citizens by shooting or locking them in barns and churches and setting the buildings on fire.[55] The town was utterly destroyed, and de Gaulle ordered that it not be rebuilt on the same site but rather serve as a memorial.

The Jedburgh operation benefited heavily from both active and passive resistance, as well as a vast communication system. The Maquis had prearranged signals for the Allied pilots, indicating

52. Gordon A. Harrison, *Cross-Channel Attack.* (Washington: Office of the Chief of Military History, Dept. of the Army, 1951), 198-99.
53. Lewis, "Jedburgh Team Operations in Support of the 12th Army Group, August 1944," 36.
54. Schoenbrun, *Soldiers of the Night*, 259-60.
55. Ibid, 376-78.

that it was safe to drop, whether supplies or Jedburgh teams.[56] A network of Resistance members passed along information between the SOE, BCRA, and then FFI or another small group. The simple act of delivering a box of ammunition meant several people risking their lives. For the Jedburgh, a green light on the plane signaled him to jump with an enormous amount of equipment toward three burning fires, hoping to land with the greeting of friendly Maquis. Jedburghs shared a common fear of their drop zone being compromised and their greeting being from the barrels of German guns. Although most drops went according to plan, not every Maquis reception committee represented an ideal situation.

Guerilla force leadership is haphazard by nature. The lack of a chain of command or Table of Organization leads to power struggles, operational strife, differences in goals, and divided loyalties. Maquis groups often disagreed on when and where to strike at the Germans or when to hold back attacks in fear of retribution against innocent citizens. In particular, sharp divisions developed among a number of Resistance members who considered themselves either Communists or Gaullists. While unified in defeating Nazi fascism, the power struggle for control of a victorious France remained a very real concern. Who would get the credit for defeating the Nazis?[57] The FTP, de Gaulle, Britain, the Allies? Jedburghs commonly worked with multiple Maquis groups representing not only multiple political affiliations but also different geographic areas. One group might prefer a different strategy for their Department or be unwilling to fight in a neighboring Department. Team Alexander reported that their "mission had been doing all it could to get the two regiments to work together." That changed when a few leaders put aside their differences and "began to tu-toi each other," meaning use of the familiar French word for "you" rather

---

56 Singlaub, *Hazardous Duty*, 44.
57. Ibid, 48.

than the formal "vous."[58] In essence, a friendly or perhaps fraternal tone helped the Jedburghs and Maquis work as a cohesive unit.

In addition to intelligence gathering, the work of these groups, whether with Jedburghs or without them, involved hundreds of forms of sabotage. Simple sabotage involved cutting communication lines, filling German fuel tanks with sugar, or deliberate acts of manufacturing errors by those in forced labor. The OSS published a manual of simple sabotage examples and instructions and had them printed on leaflets and distributed among the populace.[59] In addition to the examples above, spreading rumors about Adolf Hitler involving his supposed Jewish heritage or sexual proclivities was encouraged in an effort to hurt German morale. Leadership in London whether OSS, SOE, or BCRA, attempted to honor requests for supplies when practical. The highly sought Sten gun was the staple of Maquis equipment, but training for all forms of sabotage proved invaluable.

Resistance members specialized in railroad sabotage. Introducing a foreign element or otherwise compromising the engine was one mode of railway sabotage, but the manufacture of incendiaries from rudimentary components to blow tracks was by far the most common mode. This wreaked havoc on German military supply lines. The delivery of supplies is a major factor for military commanders planning campaigns. Resistance railroad sabotage was so effective that German commanders could depend less and less on the reliability of timely supply deliveries. Another major problem for the Germans was the fear of ambush. Ambushes became so common that German units remained in large numbers rather than leaving small units susceptible to attack. By 1944, massive areas of France were absent of German troops, enabling easier mobility for Resistance members.[60]

58. "Report of Jedburgh Team Alexander," digital image, Operation Jedburgh, *http://www. operationjedburgh.com* (accessed August 9, 2010).
59. Bull, *Special Ops, 1939-1945*, 95.
60. Lewis, "Jedburgh Team Operations in Support of the 12th Army Group, August 1944," 20.

Invasion and Liberation

THE AXIS POWERS occupied much of western Europe by the
summer of 1944 and had constructed an enormous defensive line
known as the Atlantic Wall. The Pas-de-Calais was the obvious
choice for an invasion point, and the Allies avoided it for that reason. Although heavily defended, the northwest coast of France
provided numerous advantages. First, England was just across the
English Channel, furnishing a huge marshaling area for troops
and supplies, where training could take place free of attack, and
aircraft could launch pre-invasion bombing as well as delivery of
paratroopers and glider troops. Second, the beaches of Normandy
had the minimum necessary accessibility for troop and equipment
landings. Although no port existed, temporary harbors would suf-
fice until the port of Cherbourg and the Cotentin Peninsula came
under Allied control. Third, the French Resistance could support
the effort to some extent with its unique knowledge of enemy
troop strength, gun emplacements, purposely-flooded fields, and
vital transportation arteries such as bridges and roads. Finally,
once the Normandy region was under Allied control, the advance
toward Germany, along with Allied advances from Italy and Russia,
formed what was essentially an enormous pincer movement to
force Germany's surrender.

Operation Fortitude, the Allied plan to deceive the Germans
on when and where the invasion would take place, enhanced the
lack of coherent planning by the Germans. The scope and breadth
of the operation forced the Germans to consider invasions from
nearly every point other than the east, but compounding the
problem for the Germans was a lack of planning, material, and
authority.[61] The decision to invade at Normandy did not mean
there were not numerous obstacles to overcome. The failure of

61. Hans Speidel, *Invasion 1944; Rommel and the Normandy Campaign.* (Westport, CT:
Greenwood Press, 1971), 50-51.

the raid at Dieppe in 1942 was a painful reminder of the danger of amphibious landings. Landing on a beach means contending with tides, swells, and weather for which predictions are by no means an exact science. Allied divers performed reconnaissance on the Normandy beaches discovering mine placements, elevations, German defensive positions, and the sturdiness of the sand for supporting heavy equipment. Mines presented an enormous challenge, with very few of them disabled prior to the invasion. Pre-invasion bombing destroyed some of them but most would require diffusing by engineers on the beach after the initial invasion. By far, the greatest obstacle at Normandy was the need for a direct, daylight assault against a determined, well-entrenched enemy. The Allied command knew that there would be high casualties and an enormous logistical challenge.

The allocation of the exact quantities of troops, equipment, and supplies needed continually changed due to intelligence reports, weather, German movements, and other events in the theatre. The communications between the French Resistance and London vastly enhanced the potential for success. French citizens were vital in providing all sorts of intelligence, but German troops and agents closely monitored their activities. The German occupation included familiarity with the citizens in a given town and scrutiny of any newcomers or visitors. The summer of 1944 proved to be extremely active for the French Resistance. A series of BBC coded messages on June 4, 1944, informed the Resistance to activate four plans labeled Green, Blue, Violet, and Tortoise involving the destruction of German controlled rail, communication, power, and transportation lines.[62] In addition, the messages prepared citizens for the pre-invasion bombing that would devastate coastal communities.

The Resistance actively engaged in sabotage operations in the sixty days following D-Day, working with Jedburghs and

62. Schoenbrun, *Soldiers of the Night,* 357.

conventional forces throughout France. Operation Dragoon—the invasion by Allied forces in Southern France—required the same preparations and assistance by Resistance forces. Prior to the U.S. Army push toward Brittany, *"Le chapeau de Napoléon est-il toujours à Perros-Guirrec,"* broadcast over the BBC. The message—meaning, "Is Napoleon's Hat still at Perros-Guirec,"—meant nothing to the Germans, but told the Resistance to commit all out sabotage activities. While previous activities for the Resistance were somewhat constrained or precise, this signal meant that the Allies were attempting to break out from the Normandy lodgment and the Resistance needed to inflict maximum chaos on the German Army.

As the Allies marched through France, victorious Resistance members welcomed them and in some cases escorted them toward Paris. Lieutenant General George Patton ordered the Maquis to keep their arms and serve with the Third Army as guides and interpreters.[63] Allied leaders praised the work of the Resistance in keeping several German divisions occupied that otherwise would have been fighting conventional forces. American Lieutenant General Lucien Truscott said, "We expected a good deal of assistance from them, and we were not disappointed. Their knowledge of the country, of enemy dispositions and movements was invaluable, and their fighting ability was extraordinary."[64] Many Jedburgh veterans echoed Truscott's accolades. While there were some complaints regarding the internal strife among the Maquis, Jedburghs never complained about their courage, ability, or willingness to engage the enemy.

The compelling tale of the French Resistance includes hundreds of stories of personal devotion to country that were instrumental in the Allied war effort. While numerous other groups of guerilla fighters are no less engaging, the French Resistance included members of all ages and occupations performing acts

63. Irwin, *The Jedburghs*, 117.
64. *Joint Special Operations University and Office of Strategic Services (OSS) Society Symposium: Irregular Warfare and the OSS Model*, Report, 24-25.

of passive and active resistance at great peril. They engaged the German Army in many ingenious ways, supported the Allied war effort with limited resources, and took an active part in liberating the country from Nazi tyranny. The value of organized forces working closely with indigenous forces proved a precedent for future military operations. On July 31, 1944, Captain Aaron Bank and Team Packard dropped from a darkened sky into the South of France to fight with some of these brave soldiers.

Chapter 4

TEAM PACKARD

BY AUGUST 1944, many commanders felt that Germany would lose the war, but the question of how and when remained. The Allied breakout from the lodgment at Normandy spread through much of northwest France. For more than one year, the fighting on the Eastern Front slowly edged toward Germany. In Italy, the Allies liberated Rome, but competing interests differed on what to do after that. Winston Churchill advocated for Allied forces in Italy to travel to the Balkans, liberate Eastern European countries, and squeeze Germany from a third direction. Moreover, he feared that if the Soviet Union liberated most of Eastern Europe, they would claim these territories, spreading their brand of Communism. For that very reason, Soviet leader Josef Stalin urged the Allies to use their troops in Italy for a southern France invasion. Post-war interests aside, President Roosevelt and SHAEF Commander Dwight Eisenhower pulled resources from Italy for an invasion of southern France, known as Operation Dragoon. The invasion force—originally planned to coincide with the Normandy invasion—landed August 15, 1944, forcing the German army to spread over much of France. Eisenhower would later state that the invasion from the Riviera was one of the major contributing factors to the successful defeat of the German army.[65]

Team Packard and many other Jedburgh teams dropped into southern France in support of Operation Dragoon. The majority of the teams had no specific timelines for their mission but rather operated as the events unfolded, providing measured harassment of German forces and destroying escape routes for withdrawing forces. Granted, the Germans were operating in two different directions, literally and figuratively. Adolf Hitler issued orders for many units in France to fight to the death, particularly in the ports of Cherbourg, Brest, Toulon, and Marseilles. At the same time, Hitler hampered his commanders by limiting their authorization to move troops and equipment without his express permission. While defending against ongoing Allied invasions in Normandy and southern France, the Germans also concerned themselves with potential invasions in northern France and eastern France via Italy. The mass confusion of the German army created a fertile field for sabotage by the Jedburghs.

An American and Two Frenchmen

AARON BANK FELT confident of his team and the training they received. Many Jedburghs expressed concern that they would miss the action if dropped too late. At Milton Hall, the Jedburghs observed one team after another receiving their orders. They continued to train but focused more on team unity exercises. Team Packard's leader, Aaron Bank, selected Henri Denis after several weeks of "courting." The team's first W/T operator had communication problems with Henri, so a French operator named Jean joined Team Packard. The French soldiers rarely divulged their actual surnames (see Fig.1) fearing reprisals against their families, so much of the unclassified communication for Team Packard uses the names Bank, Henri, and Jean. There were code names assigned to each team member, but they rarely used them. Few other teams had one

---

65. William B. Breuer, *Operation Dragoon: The Allied Invasion of the South of France* (Novato, CA: Presidio Press, 1987), 247.

American and two Frenchmen, and due to Henri's limited English, the team communicated almost exclusively in French.[66] Henri and Jean's backgrounds are vague other than that they were veterans of the French regular army and that Henri's military experience had been in Senegal and Algiers.[67] Throughout their time together, Bank never suggested any problems with his team members.

Nom de Guerre	Actual Name	Rank	Code Name	Nationality
Aaron Bank	Aaron Bank	Capt.	Chechwan	American
Henri C. Boineau	Henri Denis	Lt.	Fukien	French
Jean	Marcel F. Montfort	2nd. Lt.	Formosa	French

Fig. 1

Several teams ordered for missions in the south of France, traveled by ship to Algiers for staging and preparation. Team Packard left the luxury of Milton Hall for less than stellar accommodations at their base near Blida Airport in Algiers. For nearly four months, Team Packard remained anxious for the call to jump into action. Their mission called for support of the Isotrope Military Mission in the Gard and Lozère departments in the Rhône Valley region of southern France (see Map 1 / Appendix B). Isotrope, a name rarely used other than in team reports, represented the operational name for the organization of French Resistance forces, their cooperation with one another, and coordination with Allied forces. Packard's mission entailed assisting Isotrope "in every possible way, but, in particular, were to raise and arm guerilla bands for subsequent action on enemy lines of communication."[68] While that was the specific mission, Jedburghs had great latitude to operate in a wide array of activities.

Jedburghs jumped into France with a great amount of equipment. Strapped to their torso, head, and every appendage were

66. Bank, *From OSS to Green Berets*, 19.
67. Ibid, 23.
68. Records of the OSS. RG 226. Entry 103. Box 3. Folder 78, Team Packard. National Archives. College Park, MD

items such as binoculars, .45 pistol, ammunition, carbine, musette bag, canteen, toiletries, flashlight, codebooks, and many other items all enclosed in an oversized jumpsuit.[69] Some of the maps were on scarfs tied around their necks and they had an ingenious compass disguised as a button sewn into their jackets. Each officer carried 100,000 French francs and 50 American dollars while the enlisted men carried about half that amount. The purpose of this was for general expenses but certainly to purchase anything they needed for the mission. The W/T unit traveled under a different canopy and was the most precious of their cargo. Another important element on their person was a note signed by General Marie Pierre Kœnig, commander of the FFI. The note, on SHAEF letterhead, required Resistance forces to assist the bearer of the note in things ranging from covert operations and movement to food and shelter.[70]

When the command issued the order for Team Packard to depart, Bank lay in a hospital bed recovering from a bout of fever. Bank's wife Catherine said that he often told the humorous story that the hospital was not going to release him for the mission, so he sneaked out with the help of team members Henri and Jean (as getaway drivers) and arrived at the staging area in his pajamas.[71] Their drop zone lay near the small town of Quincaille, 8 kilometers south of Florac, in the Lozère Department.[72] While briefed on the general situation in their area of operation, the leaders involved, the terrain, the enemy force's strength, etc. there were many more unknowns. The questions of Maquis infighting, equipment on hand, number of fighters at their disposal, and overall mission probability were unknowns. Their job was to parachute behind enemy lines and assess those very things. As the jumpmasters told them to get ready, their first experience as an operational Jedburgh would be a blast of wind, surrounded by a dark sky, and the hope of three

69. Dreux, *No Bridges Blown*, 86.
70. Ibid, inside cover.
71. Catherine Bank, telephone interview by author, August 18, 2010
72. Records of the OSS. RG 226. Entry 103. Box 3. Folder 78, Team Packard.

small fires below to mark their landing.

Guerilla Fighters

ON JULY 31, 1944, Aaron Bank landed with a thud, slightly injuring his head, but he quickly recovered. A reception committee of more than twenty men greeted the three Jedburghs and escorted them to a safe house one mile away. Team Packard's drop, reception, and arrival of equipment went according to plan with no complications. Their orders required them to immediately contact the Isotrope Mission leader and assist the Mission with "their technical experience, leadership, and means of communication." [73] Aware of conflicts between communist and non-communist Resistance groups, Bank knew that one of the main struggles would be the distribution of the arms and other supplies delivered by Team Packard. Isotrope had already arranged for the FTP to receive the initial drop of arms, and the next drop would go to the Resistance group of Bank's discretion. While he willingly complied, this hampered his ability to form good relations with each group.

Bank and the team stayed with the Isotrope command post for several days and met the area FTP chief, known only as Mr. Barry, and FFI chief, Major Audibert. Both groups had operated in the Gard and Lozère Departments with no overall plan, so Bank devised one for sabotage and guerilla activities. The plan called for the destruction of strategic rail and road bridges and the blocking of two main tunnels. In addition, various power and communication lines were disabled.[74] A local Nazi collaborator's vehicle was "requisitioned," and a local Resistance member volunteered to drive the team throughout the operational area so Bank and Henri could direct and inspect the activities. During the first two weeks after Team Packard's arrival, Bank ran a training school for Resistance officers, teaching them basic tactical skills and sabotage techniques.

73. Ibid
74. Ibid.

Bank managed to overcome elements of hostility toward him and work amicably with each group. Hugh Tovar, a fellow OSS officer that served with Bank in Indochina, describes him as "tough and business-like," but at the same time, he presented a very favorable impression.[75] Many others offer similar descriptions of Bank as someone that expressed the proper mood at the proper time, a valuable trait for leading guerillas. Bank shared glowing accolades for Commandant Bonbix, Major Audibert's military chief, and later recommended Bonbix for commendation in his after-action report, stating, "He constantly displayed extreme bravery and excellent leadership."[76]

As with other Jedburgh teams, Packard's initial W/T transmissions to SPOC (Special Projects Operations Center) requested supplies and in particular, arms. The SPOC was a command center for the Jedburgh teams run by the British, and it directly coordinated Isotrope. The August 3 transmission noted 3,000 unarmed men at their disposal, implying that they were useless without arms. The next several transmissions reported success with the mining and demolition operations as well as numbers and movements of Maquis troops. Bank's patience had waned by August 16, when SPOC received the message that "Mission has neglected arming a splendid Maquis. Want brens, rifles, gammons, explosives, grenades for them. Will cooperate fully but need arms [sic]."[77] This was a common theme among Jedburgh teams. Not only had they been denied needed arms, but their drop occurred far too late to fully organize and train the indigenous forces. They met cooperative and willing Maquis fighters but lacked the time and resources to employ them effectively.

Southern D-Day occurred on August 15, two days earlier than Bank had previously been told. A few days later, the Germans began retreating from the city of Alès, just south of Bank's command post

75. Hugh Tovar, e-mail interview by author, November 8, 2010.
76. Records of the OSS. RG 226. Entry 103. Box 3. Folder 78, Team Packard.
77. Ibid.

(see Map 2 / Appendix B). Resistance forces and Team Packard moved into the city and set up a command post at the city hall. Resistance leaders commonly raced to city hall to set up command in liberated cities, hoping that their political affiliation would garner power in a post war environment. For the next two days, Bank and Maquis leaders defended the city from German counter attacks, capturing or killing hundreds of Germans. SPOC ordered Bank to impede German movements from the southern coast, north to Lyon.[78] Nimes was a vital pathway for the Allied invasion to move north and Alès had been an important town from which Germans could initiate attacks. The Germans faced numerous dilemmas. Hitler ordered them to defend the occupied territories from a fresh, well-supplied invasion force while German supplies lines were weakened from Resistance sabotage. Those same Resistance forces attacked them from the side and rear or while they attempted retreat. Alès is just one of the numerous towns that the Germans fruitlessly attempted to hold, with the odds greatly against them.

Aaron Bank earned the Bronze Star for the mission as a whole, but his actions from August 25 to 30 highlight the meritorious citation. He and Cdt. Bonbix went on several reconnaissance missions, personally scouting enemy strength while taking fire from German guns. They then directed sabotage and assaults on retreating German forces, hindering their escape from the southern invasion. While the operations were highly successful in harassing the enemy, the Maquis experienced 6 to 13 killed per day[79] After his units had already captured hundreds of enemy troops a few at a time, a force of several hundred Germans sent word to Bank, that they were willing to surrender to U.S. or British troops, but not to French Resistance. Undoubtedly, they feared their fate at the hands of avenging Resistance members quick to remember the atrocities the Germans had committed during occupation.

78. Ibid.
79. Ibid.

In reality, the German POW's held by the Maquis under Bank's control treated them under Geneva Convention protections. *Milice* and collaborators did not fare so well. Bank requested extra Allied troops to facilitate the surrender and manage the enemy prisoners. Those Allied troops Bank requested failed to arrive in time, and Bank did not have the resources to take them by force. It is uncertain whether that enemy unit surrendered elsewhere, escaped, or joined another fight.

To the northwest of Alès lie the Cérvennes mountain range that Bank hoped would be the site for their combative actions, given its excellent cover and high ground for guerilla tactics. However, the vast majority of the actions he led occurred on flat terrain, becoming very costly in direct combat with a large conventional force. The hit and run tactics used by Bank and the Maquis had the strategic advantage over a retreating force with poor peripheral protection, spread thin, and becoming less interested in a "fight to the death" as Hitler had ordered. The most difficult and bloodiest fighting occurred in the streets as they moved from one town to another. By August 30, the German retreat had left the area with only a few enemy stragglers hiding in the woods. The next day, French regular forces arrived, and Bank gave all his information to the intelligence officer at the French divisional headquarters in Avignon. Team Packard and the Isotrope forces of nearly 3,000 they led, had captured 3,300 Germans, killed or wounded approximately 1,000, and taken a great amount of stores and equipment.[80]

Vive les Américains

WHILE CELEBRATING IN Nîmes, Team Packard attended a ceremony at the city hall to the cheers of "Vive la France" and "Vive les Américains." Henri pushed Bank out onto the balcony with the crowd acknowledging his American uniform with more cheers. Bank thanked them for their courage in defeating the Germans

80. Ibid

but also in their rejection of the communists. Describing this in his memoir, he called this experience the "zenith of our mission. It was the thrill of a lifetime! I've had a few since then, but none compare."[81] The team reports rarely mention Henri or Jean, and Bank's memoir has few references to them. Bank does say that he met Henri at an OSS reunion forty years later and learned of his actual name, Henri Denis. According to the few references for them, they seemed to have done their job adequately, and Bank never suggests any problems with them. On September 20, 1944, Team Packard flew from Lyon to London.

Bank's assessment of the Packard mission revealed the beginning of what he and others would correct in the development of the U.S. Army Special Forces. First, sending the Jedburghs three months earlier would have allowed for more training time of Maquis. Second, SPOC should have been more realistic about expectations of delivering arms. The failure to provide these for the Maquis as promised damaged their trust in Bank. Third, there were far too many mission leaders such as SPOC, Isotrope, FTP, FFI, Jedburghs, etc. Although they shared a common goal, competing methods to realize those goals as well as the assignment of credit for particular victories caused dissension. Bank suspected the FTP had been deceptive regarding their stores of arms and that they had an ulterior motive following the invasion. He recommended their immediate disarmament. [82]

Bank and a few others were the reception committee for Team Minaret led by Major Lancelot Hartely-Sharpe, which operated near Ganges, France with similar results and successes as Team Packard. Team Collodion, commanded by Captain Harold Hall, joined Team Quinine on the opposite side of the Cérvennes mountain range from Packard.[83] What effect, if any, did Jedburgh teams Packard, Minaret, and Collodion have in support of Operation Dragoon in

81. Bank, *From OSS to Green Berets*, 58-59.
82. Records of the OSS. RG 226. Entry 103. Box 3. Folder 78, Team Packard.
83. Ford, *Steel from the Sky*, 120, 157.

and around the Cérvennes mountain range? It is difficult to say. On a grand scale, the answer is probably minimal. Comprehensive works on World War II rarely mention the Jedburgh teams, but that is likely because there were so few of them operating for such a short time. However, as a force multiplier, they achieved great success considering the little time and few supplies they had. According to a captured collaborator, the Gestapo had built a dossier on Bank and Henri, labeling them local terrorists.[84] They definitely had some impact.

Regarding Team Packard specifically, they removed approximately 4,300 German troops from the war, secured a large amount of enemy equipment, and trained many members of the French Resistance to more effectively carry out their work. Broadly, they delayed or stopped untold Germans from an effective response to the Dragoon invasion, hindered their retreat, and distracted them from supporting other Wehrmacht units fighting the Allied breakout at Normandy. A few highly trained soldiers led a willing and motivated partisan force to combat a heavily armed conventional force. The Jedburgh teams employed indigenous forces in a productive manner with tangible results that would influence a major branch of unconventional warfare and eight years later would become a permanent fixture in the U.S. Army. As a proven guerilla leader, Aaron Bank was eager for his next assignment.

---

84. Bank, *From OSS to Green Berets*, 50.

Chapter 5

OSS OPERATOR

BANK SPENT THE next several months waiting for his next mission. The Jedburgh program ended, not for lack of success, but for
lack of need at this stage of the war. OSS operators shared rumors
that their next missions would be in Asia while the need for covert
operations in Germany remained uncertain. Bank requested a
mission in Germany, reminding mission planners of his fluency in
German. Some of the missions involved parachuting alone, deep
into Germany, assessing various installations while impersonating
a German officer, and then making his way back to Allied lines
with vital intelligence.[85] Most of them involved unnecessary risks
and lacked full vetting by operational staff. After the Battle of the
Bulge, the Allied advance was outrunning unconventional mission
planners. Bank took leave in Florida, fully expecting a mission in Asia
upon his return to OSS headquarters in London in January 1945.

During the first few months of 1945, the German army was
in full retreat and soon had to defend its own soil on both fronts.
The political concerns for the occupation of Germany became as
important as the military concerns. Some intelligence indicated
that the German high command, including Hitler, would make
a final stand at a redoubt in the Alps. Due to uncertainty in the
German resolve, every possibility required consideration. Indeed,

85. Bank, *From OSS to Green Berets*, 71

the Germans created a resistance force, called Werwolf that would operate in occupied German territory and be controlled from the redoubt. Hitler's mental deterioration became evident to those working closely with him as they lived in an underground bunker in Berlin. Witnesses believed reality finally set in for him at the end, and he even allowed non-essential personnel to make their way to the west, fearing retaliatory actions by the Russians. SHAEF, not knowing exactly what Hitler was doing, continued to move troops toward Berlin with thousands of German soldiers surrendering along the way. Bank returned from his stateside leave and received the mission of a lifetime.

Tell Bank to Get Hitler

THE FRENCH CONSIDERED Henri Marchand a collaborator. A French citizen from the Caribbean island of Martinique, Marchand worked as a medical gymnast (known today as a physical therapist). Born November 23, 1902, his interests were sports and medicine, although his father wanted him to become an accountant. Due to his father's premature death, medical school was not possible, and he worked at numerous upscale spas. When Germany occupied France, he moved to Marseille looking for work with German clients and eventually went to Biarritz by assignment of the German labor office. When the French labeled him a collaborator, he turned pro-German and joined the Wehrmacht.[86] Henri Marchand was actually Aaron Bank. The cover story—very similar to Bank's past and current knowledge of language and occupation—explained why a German soldier would be in Germany with that accent. When he was speaking French or German, most Germans would dismiss his poor accent since he was from the Caribbean. If asked about his previous occupation or past, he could convincingly talk about work in spas in Biarritz and the Bahamas or his father's death. The

---

86. Ibid, images of original cover story for Aaron Bank for Iron Cross mission between pages 62 and 63.

cover story would work perfectly for Bank's next mission.

The mission, known as Iron Cross, planned for the capture of numerous high- ranking Nazi officers including Adolf Hitler. Bank would command a small unit of soldiers disguised as Wehrmacht to drop into an area near the Alpine Redoubt with the authority to conduct sabotage operations if opportunities presented themselves. Capturing these officials would rob the Germans of any centralized command and hasten surrender. A few other officers would join Bank but the majority of the unit would be German POWs who held anti-Nazi sympathies or wanted to help bring the war to an end. The mission training—conducted under the authority of the OSS Special Operations branch in Paris—meant another change of station for Bank. He could not have asked for a more perfect mission. Trickling down through the chain of command, William Donovan's approval of the mission came in the form of a simple directive—"Tell Bank to get Hitler."[87]

The recruitment for Iron Cross involved extensive interviews regarding a potential member's political or ideological feelings rather than combat experience. Only Bank and a few others knew the exact nature of the mission, and the trainees only knew they might have to fight and kill German soldiers. Most rejected the idea of killing a fellow Wehrmacht soldier but had no problem killing SS troops. Eventually, Bank and his team selected approximately 170, with most being communists and deeply opposed to fascism. Their training took place just outside of Paris in an atmosphere much like Milton Hall. Since they were posing as a German mountain company, they trained in German uniforms and with German weapons, and they spoke German. The soldiers, well screened by Bank and the other officers, trained continuously for the mission but were free to move about with no prison guards.

In April 1945, the unit was ready to drop into the Alps, with

87. Ibid, 92. Bank's superiors reportedly told him of the meeting with Donovan in which the OSS chief wanted to ensure Bank was the correct officer for the mission. Bank was fond of this quotation and repeated it in numerous interviews.

weather canceling several departures. Then the U.S. 7th Army moved into the Alpine area and Iron Cross was canceled, deeply disappointing Bank and other mission members. Had Iron Cross moved forward, it would probably have failed to achieve any significant results since the intelligence regarding the Alpine redoubt turned out to be false and many of the targets for Iron Cross committed suicide. It would, nevertheless, have been the most exciting mission of Bank's life, and it was one about which he wrote a novel with E.M. Nathanson, author of *The Dirty Dozen*. The book, called *Knight's Cross*, is a fictional account that starts in December of 1944 with the mission moving forward. The novel makes for interesting reading, but with Bank's statement that he will not say "when fictional characters take over his story," it lacks the validity for a good source on Iron Cross.[88]

The Indochina Maelstrom

JAPAN OCCUPIED THE French colony of Indochina during World War II, allowing the Vichy government to remain in nominal control. The Japanese imprisoned both Vichy and non-Vichy French officers while others evaded capture and recruited local inhabitants for guerilla warfare, primarily in Laos. The heavier Japanese presence in Vietnam prevented operations similar to those in Laos, but Ho Chi Minh formed the communist Viet Minh, seeking independence from French control. During the tumultuous years of World War II, Indochina experienced numerous outside forces seeking control of the area, such as Japan, France, and China while various local groups sought independence. The Potsdam Conference failed to give Charles de Gualle proper authority for reclaiming French Indochina.[89] By the end of the war, China controlled Indochina above of the 16th parallel, and the British below

---

88. E. M. Nathanson and Aaron Bank, *Knight's Cross: A Novel* (New York: Carol Pub. Group, 1993), vii.
89. Arthur J. Dommen and George W. Dalley, "The OSS in Laos: The 1945 Raven Mission and American Policy," *Journal of Southeast Asian Studies* 22, no. 02 (September 1991): 329.

it, but mainly for the purposes of facilitating the Japanese surrender. The OSS office in Kunming, China concentrated its efforts on the Viet Minh throughout Indochina but its primary concern for them was in the north.[90]

Aaron Bank reported to the OSS Special Intelligence branch in China in July 1945. Fitness reports commend his leadership, judgment, and level-headedness making him an excellent candidate for peacetime intelligence work.[91] A project developed called the Bank Mission, in which Bank would lead 125 French and Annamite (previous name for Vietnamese) soldiers and men on an intelligence mission in northern Vietnam. Operating as a guerilla unit, they would raid, ambush, and harass Japanese troops and garner any intelligence possible from prisoners.[92] Unfortunately, Bank again experienced something he called the abort jinx, when the mission was canceled due to concern over potential Viet Minh hostilities with the French soldiers serving with Bank. Within a few days, atomic bombs dropped on Hiroshima and Nagasaki. Bank then left for a short trip to Hanoi, investigating Japanese war crimes and making assessments of prisoners of war.

Because it was a French colony, many of the indigenous peoples of Indochina spoke French, making Bank a good candidate for operations there. His next assignment, known as the Raven Mission, was primarily a POW mercy mission. Several missions similar to this were carried out in Asia, including one that found four airmen from the Doolittle raid and another that rescued General Jonathan Wainwright.[93] Bank's team would parachute into Vientiane, Laos (see Map 3 / Appendix B), search for a prisoner of war camp in that area, and then provide food, and medicine and facilitate the evacuation of the prisoners. The mission eventually evolved into

90. Smith, *OSS: The Secret History*, 309.
91. Records of the OSS. RG 226. Entry 224. Box 34. Personal File: Bank, Aaron.
92. Records of the OSS. RG 226. Entry 154. Box 199. Field Station Files: Kunming. National Archives. College Park, MD.
93. Records of the OSS. RG 226. Folder 3371. Box 198. Field Station Files: Kunming. National Archives. College Park, MD.

international relations as with many OSS and later CIA missions in Indochina, where communist tendencies were brewing. The team consisted of seven officers and two enlisted men. The biggest unknown for the team would be the level of hostilities between British, French, Annamites, Chinese, and even remnants of the Japanese forces.

Team Raven successfully landed on September 16, 1945 at Vientiane, although one enlisted man suffered a fractured foot, and 35% of their dropped supplies received damage due to malfunctioning parachutes. Although they found no POW camps, they did help over a hundred refugees with humanitarian aid and evacuation. Most of the Japanese had left except for those with business interests, and Team Raven met little or no resistance. However, their largest concern became the extremely tenuous situation between Annamites and French colonial soldiers. The French adamantly asserted their authority and right to reclaim their colonial territory that the Japanese had occupied. The POW mercy mission developed into a diplomatic mission with Team Raven members caught in the crossfire.

Three days into the mission, Bank and two other officers went to Udon Thani, Siam and Thakhek, Laos to seek information on POWs and plan for potential evacuation scenarios as well as discuss the hostile French-Annamite situation with various parties. One of those parties was Viet Minh leader Ho Chi Minh. When their transportation fell through, Ho gave Bank a ride to Thakhek, discussing various political situations along the way. He expressed to Bank that he felt that, unlike the Europeans, the Americans were not colonizers; he cited the independence given to Cuba and in process in the Philippines as examples of his point. Ho hoped that the United States would give the Viet Minh economic, financial, and military support for their independence.[94] Knowing the perception of Ho, since he had sought education in China and particularly Marxism

94. Bank, *From OSS to Green Berets*, 118.

in Russia, Bank expressed to many that Ho Chi Minh was "not a communist yet, but if we don't help, he will talk to the Russians and they will help him."[95] Listening to the concerns and requests of indigenous peoples would become a major tenet of U.S. Army Special Forces procedures. Leaving the meeting with Ho, Bank and other Team Raven members spent several days assessing the various local leaders and hearing of their situations. In addition, they continually urged French leaders to stand down. The French continued to suggest that if their territorial demands went unmet, there would be bloodshed and they had no intention of abiding by any Chinese authority. Defeating the Annamites was France's first step toward regaining Indochina.[96]

Thus, the belligerents during this time were the Annamites, consisting of Viet Minh and Free Laotian forces, against French colonial soldiers, with the British providing some non-combat support to the latter. Bank reported to headquarters that the British were operating above the 16th parallel in opposition to Potsdam and requested that Chinese troops disarm the French.[97] The Annamites were poorly armed, with some fighting only with bamboo spears while the French had an abundance of arms, some of them American made supplied to them the by the British. While Team Raven experienced no actual combat, various episodes of street fighting occurred around them and in one instance, the local governor of Thakhek told Bank that his city was "under siege."[98]

A major incident occurred during the mission that provides a prime example of the hostile situation during this time and various ruling authorities. One Team Raven member, 1st Lt. Algar Ellis, planned a trip from Siam to Thakhet, via boat. Two other officers on an unrelated matter, British SOE Major Kemp and French Lt. Klotz, shared the boat with him. Ellis warned them that any

95. Catherine Bank, telephone interview by author, August 18, 2010.
96. Records of the OSS. RG 226. Entry 154. Box 174. Field Station Files: Kunming. National Archives. College Park, MD.
97. Records of the OSS. RG 226. Entry 154. Box 174. Field Station Files: Kunming.
98. Records of the OSS. RG 226. Entry 154. Box 174. Field Station Files: Kunming.

French officer in Thakhet could expect trouble. While they were ascending the steps in Thakhet, an Annamite patrol stopped them. Ellis stood several feet back since they did not turn their attention to him. The conversation was in French, and Ellis, who spoke no French, suggests it was very heated and it appeared the Annamites wanted to arrest the French officer. Kemp and Klotz started walking back to the boat when the patrol shot and killed Klotz. Bank met with local leaders and thoroughly investigated the matter. The two Annamites accused of the murder of Klotz claimed that he and Kemp were running away. Bank trusted the Laotian leaders to deal with the two men according to the regulations of the Chinese army.[99] The Annamite patrol considered this a justified shooting, the British a murder, and Bank an unfortunate situation resulting from Klotz's own actions.

Based on this incident and skirmishes in several cities in Indochina, Team Raven left for Siam, and after three weeks in the field, received orders to return to Kunming. Regarding success, Team Raven saw few tangible results. Team Raven member Hugh Tovar agreed that it was "hard to say" if they were successful and felt the murder of Lt. Klotz was blamed on Team Raven.[100] Bank earned the Soldier's Medal for two main reasons. First, he volunteered for the mission in an area thought to be under Japanese control. Second, in spite of the "explosive" situation, he moved about the area with "great tact, diplomacy, and courage."[101] Bank felt that the ability of the OSS Special Intelligence branch to support them was severely lacking. The main purpose for the Team Raven mission had been to search for and assist in the evacuation of innocent internees where possible, and in this they had been successful. In doing this, they exerted certain diplomatic authority for matters of safety, but could do little to dissuade the right of French sovereignty in Indochina. Some authors argue that President Harry Truman

99. Ibid.
100. Hugh Tovar, e-mail interview by author, November 8, 2010.
101. Records of the OSS. RG 226. Entry 92a. Box 87. Field Station Files: Kunming. National Archives. College Park, MD.

and his successors, used the Raven mission as a reason for tighter control over the OSS's successor, the CIA.[102]

OSS Phase-Out

PRESIDENT TRUMAN DISBANDED the OSS on September 20, 1945. While Roosevelt had championed the organization, Truman and others, such as FBI Director J. Edgar Hoover, resented the clandestine intelligence network.[103] OSS members dispersed throughout various other agencies, but many would come back together with the formation of the CIA in 1947, an agency very much like the Special Intelligence branch of the OSS. The OSS influence on special warfare operators continues to this day for various reasons but following World War II, the OSS model focused more on people than technology. Many combatants throughout the world shared relatively similar technology, but the ability to employ soldiers or other agents in relational or force-multiplying combat roles needed specific organizing.[104]

Following the disbanding of the OSS, Aaron Bank served in various U.S. Army administrative positions such as the Strategic Services Unit in the Office of the Assistant Secretary of War and the Counterintelligence Corps. During this period, he devoted himself to the Army's formal adoption of an unconventional warfare program and intensely studied the history of unconventional war. While the CIA adequately addressed intelligence concerns, Bank felt the OSS Special Operations branch would fit perfectly in the Army. While serving as executive officer with the 187th Regimental Combat Team during the Korean War, he closely witnessed the operations of conventional forces. Bank's daughter stated that while her father's OSS experiences certainly equipped him with knowledge and experience in unconventional war, the inefficient use of soldiers in Korea was the real catalyst for his belief that the

102. Dommen, "The OSS in Laos," 346.
103. Smith, *OSS: The Secret History*, 333.
104. *Joint Special Operations University: Irregular Warfare and the OSS Model.*, 13.

Army needed to create a special force.[105] In other words, he knew that unconventional forces engaging in reconnaissance or tactical operations could have saved many lives. In 1951, a transfer order sent Bank from a combat role to a desk job with the Army Psychological Warfare staff, but he had no idea why.[106]

105. Linda Ballantine, telephone interview by author, August 18, 2010.
106. Bank, *From OSS to Green Berets*, 139.

Chapter 6

A SPECIAL FORCE

WHILE FASCISM AND militarism had been the main concern for leaders in the West during the war, communism emerged as the primary postwar threat. The Allies met for conferences at Yalta and Potsdam in 1945 seeking to bring agreement on postwar organization of countries previously occupied by Germany as well as the occupation of Germany. Tensions emerged over differences for the type and amount of control that countries such as the United States, Great Britain, and the Soviet Union should have. In particular, there were concerns over the Soviet Union's control of those countries that would eventually become the Eastern Bloc, such as Czechoslovakia—countries liberated from the Nazi boot only to bow in submission to the Hammer and Sickle. As the Cold War ramped up, military planners considered Eastern Europe a likely potential region for future combat operations. Knowing that there would likely be thousands of indigenous citizens opposed to communism in each of the Eastern Bloc countries, consideration of how to harness those citizens as a guerilla force spurred serious thought.

Having successfully defeated the Axis powers with conventional forces, the U.S. Army command gave little consideration to those seeking a permanent unconventional force. Although many acknowledged the contributions of unconventional forces, their role did not seem very significant, and if needed could easily resume should another major conflict arise. The Army Rangers suffered from this mindset as well. Although several battalions formed early in the war and achieved admirable successes, they were quickly disbanded at the end of the war. American military leaders typically considered the Rangers an elite force rather than an unconventional one during this era, and thought units could be quickly activated should the need arise. Army planners considered the Rangers, with their quick strike actions and deep penetration capability, a valued asset but not one that required a peacetime formation. Elite and unconventional force successes only supported the notion that they could form and train once hostilities commenced.

Psychological warfare garnered more respect among the Army command even though its effects were difficult to measure. General Eisenhower, in a letter to the SHAEF Psychological Warfare Division Commander, acknowledged that the contribution of psychological warfare toward the final victory could not be measured in towns destroyed or barriers passed but that it held a place of importance in the military arsenal.[107] Psychological warfare, as well as unconventional warfare, fell under the general category of *special warfare*, which was in wide use during this period. The type of force Aaron Bank and others sought to create would find an ally and organizational partner in the Office of Psychological Warfare.

UW Veterans on a Mission

THE MAN WHO ordered the transfer of Aaron Bank from a combat assignment in Korea to the Psychological Warfare staff was the former director of the Psychological Warfare Division for the European Theatre during World War II, Brigadier General Robert McClure. Today, many know him as the Father of U.S. Army Special

---

107. Alfred H. Paddock, *U.S. Army Special Warfare: Its Origins* (Lawrence, KS: University Press of Kansas, 2002), 20.

Warfare and the U.S. Army Special Operations Command honored him by naming their headquarters after him. While many played roles in the development of U.S. Army Special Warfare, McClure led the campaign for the permanent establishment of a specific unit trained for unconventional war. In addition to Bank, McClure formed a team of World War II unconventional force veterans including some that served with Merrill's Marauders and some that led guerilla forces in the Philippine Campaign.

The one person McClure brought in prior to Bank was Colonel Russell Volckmann, a graduate of West Point. After the fall of Bataan in 1942, Volckmann, serving as an infantry instructor to the Philippine Army, organized a guerilla resistance force of 22,000 Filipinos that waged an effective war against the Japanese for three years. In reference to General Douglas MacArthur's declaration that "I shall return," Volckmann stated, "We remained."[108] Among his early contributions was FM 31-21, the guerilla warfare field manual for the U.S. Army, still in use today. His initial contributions outlining Special Forces operations became the template from which he, Bank, and McClure lobbied the higher command. Bank, meanwhile, developed the TO&E (Table of Organization and Equipment) for Special Forces loosely based on the OSS operational groups.

As previously mentioned, Bank spent the immediate postwar years studying the history of unconventional war. Similarly, Volckmann researched every major resistance movement he could in preparation for the writing of field manuals on guerilla warfare.[109] Bank, Volckmann and others widely discussed the best model for furthering the idea of a Special Forces unit. Since the proposed unit was seeking permanent status in the U.S. Army, they needed to show how prior units had functioned and how they would contribute in the warfare of the future, in particular against the growing threat of

---

108. R. W. Volckmann, *We Remained: Three Years behind the Enemy Lines in the Philippines* (New York: W.W. Norton, 1954), 220.
109. R. W. Volckmann to Mrs. Beverly E. Lindsay, March 21, 1969, Alfred Paddock Private Collection.

communism in Eastern Europe. Certainly, Ranger units had many successes, and all agreed that Special Forces units should have that level of combat training. The exploits of Darby's Rangers in Africa and Italy, Rudder's Rangers at Pont du Hoc, and Merrill's Marauders in Burma were well known. The major difference, though, is that Rangers operate on their own as a quick strike force on short-term missions whereas the Special Force units would operate in small numbers, well behind enemy lines, to organize, train, and lead indigenous fighters on a long-term basis.

Bank succinctly stated that Special Forces "have no connection with ranger type organizations since their [Special Forces] mission and operations are far more complex, time consuming, require much deeper penetration and initially are often of a strategic nature."[110] Former Jedburgh, John Singlaub had several conversations with Volckmann on the subject of the proper model at Ft. Benning in 1949. Singlaub argued that Volckmann's experience was not the proper model because his Filipino guerilla force was largely trained and organized. The French Resistance—which had little military training and organization—was a much better model of what a potential Special Forces unit would encounter.[111] Special Force soldiers would likely organize pockets of resistance in some particular Eastern European country against communist elements. In addition to combat skills, they would need to have great adaptability, leadership, and relational and organizational skills.

Ultimately, Bank and Volckmann settled on the OSS Jedburghs as the concept model and the OSS Operational Groups as the size model. Jedburgh veterans complained that they went into France too late and some suggested larger Jedburgh teams as a more effective model. The typical Operational Group's size was thirty men with the capability to split in two and operate separately. Bank's original design has changed little to this date, other than in the

110. Aaron Bank to Mrs. Beverly E. Lindsay, February 27, 1973, Alfred Paddock Private Collection.
111. John Singlaub, telephone interview by author, December 03, 2010.

number of soldiers within each unit and the specialties required. Bank's original basic unit—named a Special Forces Operational Detachment A, and now commonly called an A-Team—would work independently over a long period of time. There were two officers, a captain in command with a 1st or 2nd lieutenant as executive officer, and ten enlisted men. The enlisted men each had specialties such as medical, weapons, communications, and operations. Two men, for example, would be the medics, but every member of the unit had training in each other area.[112]

Several A Detachments formed a B Detachment and then several B Detachments formed a C Detachment. The B and C detachments held their own complement of officers and staff to oversee the underlying detachments and were assigned to a specific operational area. The B Detachment would be responsible for a specific region within a country and might have as many as 1500 indigenous guerilla fighters under its control while the C Detachment would have responsibility for an entire country. All these detachments fell under a Special Forces group with operational responsibility for a geographic region, such as Eastern Europe.

The first group, the 10th Special Forces Group, used the number *ten* hoping to deceive opponents that there were at least nine other Special Forces groups throughout the world. The 10th Special Forces Group operated under the Psychological Warfare branch. Conventional forces distanced themselves from psychological operations due to their non-combat nature. However, Psychological Warfare garnered respect within the military community. Lacking the formal honor and history of an established unit, the Special Forces operated under the Psychological Warfare branch, which held such honor and history.[113]

Before final authorization, McClure, Bank, and Volckmann experienced stiff resistance from various groups. Conventional

112. Charles M. Simpson, *Inside the Green Berets: The First Thirty Years, a History of the U.S. Army Special Forces* (Novato, CA: Presidio Press, 1983), 36.
113. Paddock, *U.S. Army Special Warfare*, 154.

forces felt that unconventional units were undisciplined and lacked training, and that they should not receive precious personnel billets and finances. The CIA felt that unconventional warfare should fall under its operations and that similar U.S. Army units would cause conflicts within foreign countries. Even the Air Force argued that they should control this type of warfare. Bank credits Volckmann with overcoming the majority of this resistance.[114] In early 1952, the army canceled thousands of Ranger personnel billets, freeing up space for the Special Forces. Army Chief of Staff, Major General J. Lawton Collins, authorized the Psychological Warfare Center at Ft. Bragg, North Carolina, which included the 10th Special Forces Group.[115] It is important to note that there were many unconventional warfare veterans who contributed to the creation of the Special Forces, and it can truly be classified as a team effort. In summary, McClure provided the rank and connections, Volckmann the master plan, and Bank the basic framework.

Inaugural Commander

MCCLURE CHOSE BANK, now a colonel, as the first commander of the 10th Special Forces Group. The first muster, and thus the official date of activation for the U.S. Army Special Forces was June 19, 1952 with Bank's report that, "Present for duty were seven enlisted men, one warrant officer, and me, making a slim morning report."[116] Recently, the question as to why McClure chose Bank rather than Volckmann as the first commander has caused some debate. Part of the reason for this debate is that Bank received the honorary title of Father of the Special Forces, largely because he was the first commander. Volckmann had undoubtedly worked tirelessly to establish the Special Forces, written two army manuals on the subject, and spent three years leading guerilla forces in the

114. Aaron Bank to Mrs. Beverly E. Lindsay, February 27, 1973, Alfred Paddock Private Collection.
115. Paddock, *U.S. Army Special Warfare*, 139.
116. Bank, *From OSS to Green Berets*, 171.

Philippines. However, the emphasis for the 10th Special Forces would be in Europe, a region where Bank had much experience. In addition, Bank had trained and operated within the OSS, the very model for the Special Forces. Finally, Volckmann had done an impeccable job of convincing the Army of the need for this force, and McClure may very well have wanted him to continue those public relations efforts. Thus, Bank's title of Father of the Special Forces was more a result of being the first group commander, rather than an attempt by anyone to give him more credit than Volckmann.

Bank wanted the best soldiers in his command but recruiting did not come easy. Army leaders were reluctant to lose good soldiers to this "renegade" new organization and dissuaded soldiers from volunteering. The pamphlets and other means used to recruit new members lacked adequate explanation. Bank asked for the TO&E to be declassified for a more thorough explanation of the new unit and to allow current members more openness to discuss their unit.[117] Bank wanted the best soldiers knowing that they must be highly disciplined individuals that could work well in small teams over a long period of time. In addition, they needed strong relationship skills since they would be working with people of different cultures than their own, including some serving in their own units. This was due to an area of recruitment for the Special Forces made possible by a new law allowing enlistment of foreign-born soldiers, in particular those from Eastern European countries.

The Lodge-Philbin Act, commonly called the Lodge Bill and promoted by Massachusetts Senator Henry Cabot Lodge, passed Congress on June 30, 1950. The bill allowed qualified applicants to serve in regular army units without segregation for a term of five years, and following that period, they could become American citizens.[118] In a speech to the first group of Lodge Act soldiers,

117. Paddock, *U.S. Army Special Warfare*, 202.
118. U.S. Congress, Senate, 1950, *Public Law 597*, S.S. 2269, 81st Cong., 2d sess., Accessed 12 Jan 2011; available from GPO Access.

Senator Lodge told them that their "knowledge of foreign languages, customs, and terrain will be of great help to this country." In addition, he wanted to assure them that they were not to feel like mercenaries, but enjoy equal standing as American soldiers.[119] As with the Special Forces, many elements opposed this new type of soldier, likening them to foreign legionnaires. Many of these soldiers would serve with distinction in the Special Forces, the most famous of them being Lauri Törni, a veteran of the German and Finnish armies. Special Forces veterans from the early years are quick to acknowledge the contributions of the "Lodge Bill Men" from their units.

Training at Ft. Bragg and nearby Camp McCall involved instruction in communications, weapons, maps, explosives, and hand-to-hand combat. Bank would occasionally interrupt an instructor to make a teaching point but was quick to affirm the instructor's leadership. In addition, he personally set the pace in many aspects of physical fitness.[120] The troops also conducted long-term exercises in Georgia, where the terrain was more suitable for mountain, swamp, and forest training. Special Forces sometimes performed missions involving contact with local citizens in which they had to use those civilians for some tactical goal. Very much like the work (known as schemes) done by the Jedburghs, these exercises were longer and extremely challenging, and they sometimes caused complaints from local law enforcement. Today, these exercises comprise one major evolution for completion of Special Forces training called Robin Sage.

In late 1953, the 10th Special Forces Group split to form a second unit, the 77th Special Forces Group. The 77th remained at Ft. Bragg while the 10th deployed to Bad Tölz, Germany. Russell Volckmann's 1954 memoir notes that "the people of Poland,

119. Henry C. Lodge, "Statement of Senator Henry Cabot Lodge" (speech, Address to First Group of Enlisted Aliens, Camp Kilmer, New Jersey, October 15, 1951), Massachusetts Historical Society, Boston.
120. Edward Fitzgerald, e-mail interview by author, May 19, 2010.

Czechoslovakia, the Ukraine, Bulgaria, and Albania will be con-
ditioned to guerilla warfare." He goes on to name several other
countries as well.[121] The implication was that after some years under
Soviet control, pockets of partisans would be ready for Special
Forces to help lead them in guerilla campaigns to resist or remove
communist rule. Although training continued in such things as lock
picking, cross-country skiing, and mountain climbing, the strategic
location of Bad Tölz meant A Detachments could easily deploy
into communist countries at any time.[122] Language training for a
soldier's particular target country commenced and in addition,
most soldiers were required to learn German and Polish. The basic operation they expected to encounter, known as a stay-behind
operation, required that Special Forces soldiers would infiltrate a
country before it was overrun by enemy forces, and then organize
a resistance movement. At present, veterans from the Bad Tölz
days will neither confirm nor deny any actual missions.

The Green Berets

BANK LEFT THE 10th in 1954, due to a normal rotation, to become
the Chief of the Plans and Operations Division in Europe for the
Seventh Army. He left the unit, proud of their accomplishments,
and confident in their operational readiness.[123] The Army turned
down his many requests for a distinctive headgear for the unit to
acknowledge their special training and purpose. Approval for the
green beret would not come for another ten years, when President
John F. Kennedy authorized its use. The average American likely
first became acquainted with the Green Berets from the 1968 John
Wayne movie called *The Green Berets*. However, a major figure in
promoting of the Special Forces was William Yarborough, a West
Point graduate and World War II airborne veteran. His enhance-
ments of the Special Warfare Center and School and his creation

121. Volckmann, *We Remained*, 231.
122. Simpson, *Inside the Green Berets*, 42.
123. Bank, *From OSS to Green Berets*, 204.

of new Special Forces groups, as well as hosting President Kennedy for a visit in 1961, grew the Special Forces substantially.[124]

The implementation of Special Forces in the Eastern European satellite countries never happened, at least not to any recognizable scale for public knowledge. The Special Forces' first major test occurred in Vietnam. Regardless of the successes and failures of the U.S. Army as a whole, Vietnam confirmed the basic tenets of Volckmann's philosophy and Bank's organization for the Special Forces. One accomplishment of Special Forces in Vietnam was the successful integration of 45,000 South Vietnamese that previously lacked leadership into the regular fighting forces of that nation. In addition, the large number of Medals of Honor awarded highlighted Special Forces soldiers' combat skills and bravery. Their legacy continues as Special Forces units lead, train, and organize tribesmen to battle the Taliban in Afghanistan.

---

124. Simpson, *Inside the Green Berets*, 66-68.

CONCLUSION

THROUGHOUT THIS THESIS, terms such as guerilla warfare, special warfare, and unconventional warfare seemingly intersect. There is difficulty in using these terms because so many definitions exist. Respected authors, military personnel, and the Department of Defense sometimes differ on these terms or omit them altogether. However, these differing terms typically do not contradict one another but only slightly vary. This thesis has attempted to use them in the context that the majority have used them. For example, this thesis frequently used the above three terms with guerilla warfare a part of special warfare, and special warfare a part of unconventional warfare. Most importantly, these terms imply operations outside the bounds of conventional warfare. To explore the process by which the U.S. Army implemented the Special Forces, the reader must understand the clear division that existed between conventional and unconventional forces regarding tactical ability, attitudes, and interpretation of effectiveness.

Prior to World War II, some of the world's militaries had variations of unconventional force units, but most of these are better classified as elite units such as some of the British commando units. The historical record, in reference to modern militaries, demonstrates that unconventional forces have always been used and been effective. That grand scale of World War II required the United States military to explore alternative methods outside standard conventional force practices. Aside from negative attitudes, proponents of World War II unconventional forces encountered several other difficulties such as the development of correct training, use of special equipment, inadequate senior organization, and senior conventional staff being unaware or unsure of their capabilities.[125] Regardless, units such as the Jedburghs persevered, and although they certainly were not the turning point in the war, they achieved their goals and only fell short of their full potential due to reasons outside their control. Shortly after the war, conventional forces could certainly boast their success, but enough senior officers saw the continued potential for unconventional forces.

While certain elements within the U.S. military may still harbor opposition to unconventional forces, they are among the minority. Every branch of the U.S. military has one or more elite or special operations units. Aside from the Special Forces and the Rangers, the U.S. Army created the 1st Special Forces Operational Detachment-Delta, or Delta Force, designed to respond to terrorism. The U.S. Navy Seals—whose legacy comes from World War II era frogmen, underwater demolition teams, and the OSS Maritime Units—are among the premier special operation units in the world. Similar to Army Rangers, the U.S. Marine Corps' Force Reconnaissance units perform deep penetration actions but typically in cooperation with conventional units. Even the U.S. Air Force and U.S. Coast Guard contain special operation units with particular specialties germane to their parent branch. Not only are unconventional units a permanent fixture in the U.S. military, foreign militaries throughout the world duplicate them with intercountry coordination for training.

Father of the Special Forces

AARON BANK'S ARMY career was anything but typical. He was not a highly decorated hero such as Audie Murphy, but he certainly exhibited bravery. He did not hold a West Point pedigree but held cosmopolitan wisdom. He entered the service twenty years older than his fellow soldiers. Very few soldiers of his era experienced

125. Lewis, *Jedburgh Team Operations*, 60.

unconventional war, but he embraced it. He met William Donovan and Ho Chi Minh. It is possible that he was one of the only officers tasked with the capture of Adolf Hitler. However, his legacy is in the organization he helped to create. Although this thesis has interwoven his biography in telling the story of the creation of the U.S. Army Special Forces, there has been no attempt to elevate Bank as more important than any other founder. Rather, Bank serves as a composite for the several men who passionately pursued this goal because of their wartime experiences.

Following Bank's service in Europe, he returned to the United States, serving in various army staff positions before retiring in 1958. Settling in Southern California, he held various residential management and security positions. One of these was head of security for a beach road development where David Eisenhower (grandson of Dwight D. Eisenhower) and Julie Nixon (daughter of Richard Nixon) lived and had received threats. Bank confronted one intruder who had likely been making the threats—a man much bigger than Bank—and restrained him until police arrived. The police and his family wondered how he managed to do this, but in keeping with his clandestine nature, he would only say, "I have my tricks."[126] Another incident where Bank chose to intervene concerned security at a nuclear generating station in San Clemente, California. Bank argued that a trained Special Forces soldier could easily sabotage the plant by himself, causing horrific events. This eventually led to testimony before a congressional committee and increased security at various plants. Interestingly, several years later, Navy Seal teams created a unit known as Red Cell to conduct mock sabotage of military installations and other vital U.S. properties to look for vulnerabilities.

Bank met his wife Catherine in Germany in 1945. A Luxembourg native and fluent in German, French, and English, Catherine worked as an interpreter between the U.S. Army and German

126. Linda Ballantine, telephone interview by author, August 18, 2010.

prisoners. One day at a base pool, Bank attempted to impress Catherine by pulling himself up out of the pool into a handstand on the pool edge using nothing but his arms. It worked and they had two daughters and remained married until his death. He continued physical fitness well into his nineties, swimming up and down in the ocean every morning regardless of temperature. Because of his civilian occupation as a realtor, he was home a lot and spent much time with his daughters, but they knew little of his past. His oldest daughter Linda found out when a friend at school showed her a book with her father's name in it. He was proud of the title of Father of the Special Forces but there remains no indication that he ever invoked the title or sought praise. Like many of his generation, he felt his service was his duty as an American.

The Force Multiplier

THE DIFFERENCES BETWEEN conventional and unconventional forces are common knowledge, but few understand why the Special Forces is different from most every other special operations unit. In the most rudimentary form, Special Forces soldiers will say that while most kick doors down, we kick doors down and stay awhile. The more elaborate definition is that most special operations units perform quick strike tactical operations for a specific objective. The Special Forces Group can do that work, but it is typically only an initial function or future function within an operation. Their role requires them to establish a relationship with local leaders, whether guerilla or partisan, reach an agreement on long-term goals, then train and/or lead them in combat or infrastructure project that will affect a military objective. A small, highly trained Special Forces team can organize a previously ineffective group and greatly multiple the force that group can apply.

The world witnessed conventional war on a scale never seen in modern times during the years of 1939-45. Unseen for the most part, were various units applying special skills in unconventional means. Some of those, the Jedburghs, used the force multiplier principle, effectively hindering the German Army's response and retreat during the Normandy invasion and breakout. One might argue that their true achievement was the theory of what they could have done if deployed in May 1944. Could they have already secured many of the needed bridges? Could they have provided a distraction for Germans forces firing down on Utah Beach? Would this have prevented the high casualty rate for the very dangerous job of paratroopers? These questions are not fruitless, and in the post-war years they fueled the fervor of Bank to promote the Special Forces as a permanent unit.

Mark Nutsch and Laval Simons know of Aaron Bank. They are former Green Berets, and when asked about Bank, they immediately acknowledge his contribution. In late 2001, Nutsch led a group of Northern Alliance guerilla fighters on horseback to fight the Taliban. Simons has served in Afghanistan as well. Now civilians, they both volunteer for an organization called ACT (Afghan Care Today) that seeks to serve the people of that country one tribe at a time. After reestablishing the relationships they built during their service, they help dig water wells, build schools, and start small businesses. Not just anyone can do this type of work due to the consistent danger of Taliban attack. As Simon explains, helping local tribes become self-sufficient and helping young men get an education or find a job threatens the Taliban. The Taliban want tribes to depend on them, and they need those young, uneducated men, with little hope, to desire a better life with the Taliban and be willing to raid and kill to accomplish their goal.[127] That is where the unique skills of ACT are used. Not only do they have strong relational skills to connect with tribal leaders, but also they can teach the local tribes to defend their water well or their newly built school. It is doubtful that Aaron Bank ever envisioned Special

127. Laval Simons, interview by author, April 17, 2010.

Force soldiers doing this type of volunteer work, but it is ideally what they have trained to do.

APPENDIX A: GLOSSARY

BCRA—Bureau Central de Renseignements et d'Action (Central Bureau of Intelligence and Operations).

Conventional Warfare—Elements of Infantry, armory, air, and naval forces organized as a cohesive forced typically engaged large scale war.

Department—Administrative/Geographical division within France.

FFI—Forces Françaises de l'Intérieur (French Forces of the Interior).

FTP—Francs-tireurs, a term describing unconventional French forces.

Guerilla Warfare—Consisting of small paramilitary elements fighting within or near a specific terrain for which those fighters might also dwell.

Irregular Warfare—Elements of warfare outside standard practice of conventional forces and used interchangeable with the more widely used special operations.

MI5—(Military Intelligence Section 5)—British unit responsible for domestic intelligence work.

MI6—(Military Intelligence Section 6)—British unit responsible for foreign intelligence work.

OSS—United States Office of Strategic Services created in 1942 for intelligence and covert operations.

Partisan Warfare—Similar to guerilla warfare but typically referring to European elements opposing Germany during World War II.

Psychological Warfare—The use of truth, partial truth, and outright deception in various media elements or by persons to confuse the enemy.

Regular Warfare—Operations conducted by conventional forces.

Resistance Force/Network—Consisting of small pockets of partisans including both combatants and non-combatants.

SOE—Special Projects Operations Center run by the British.

SPOC—British Special Operations Executive created in 1940 for espionage and resistance activities behind.

Special Forces—Lower case use of this term meaning elements of warfare outside standard practice of conventional forces and used interchangeably with the more widely used *special operations* while the upper case use of this term specifically refers to the United States Army branch also known as the Green Berets.

Special Operations—Military operations that work outside the norm of conventional forces.

Special Warfare—Elements of warfare outside standard practice of conventional forces and used interchangeable with the more widely used special operations.

Unconventional Warfare—Elements of warfare outside standard practice of conventional forces and used interchangeable with the more widely used special operations.

W/T—Wireless Telegraph

APPENDIX B: MAPS

Map 1

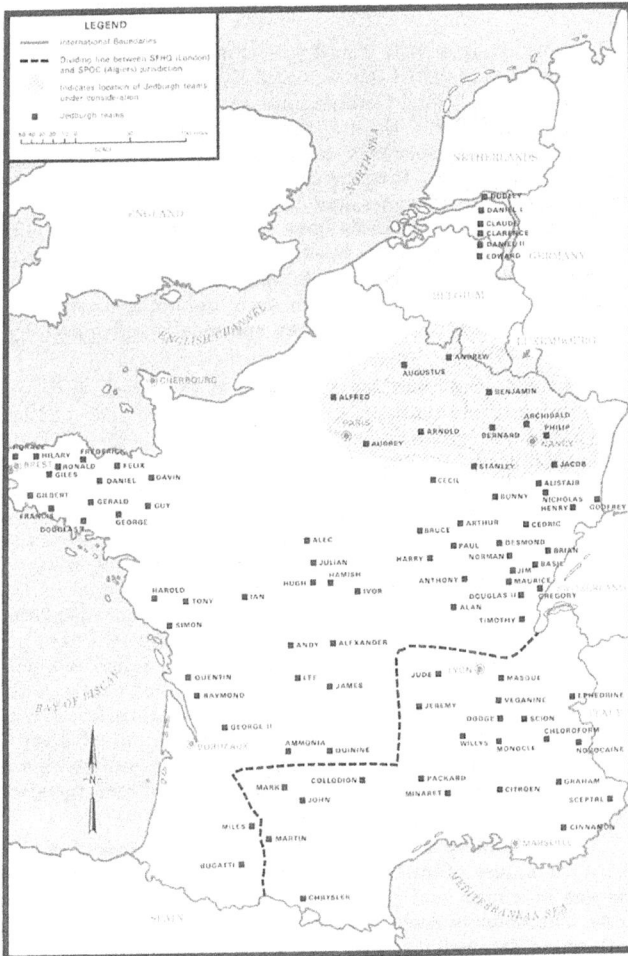

Map 1. Jedburgh teams in France, June—September 1944

Source: *http://www.cgsc.edu/carl/resources/csi/lewis/Images/pgx.gif*

Map 2

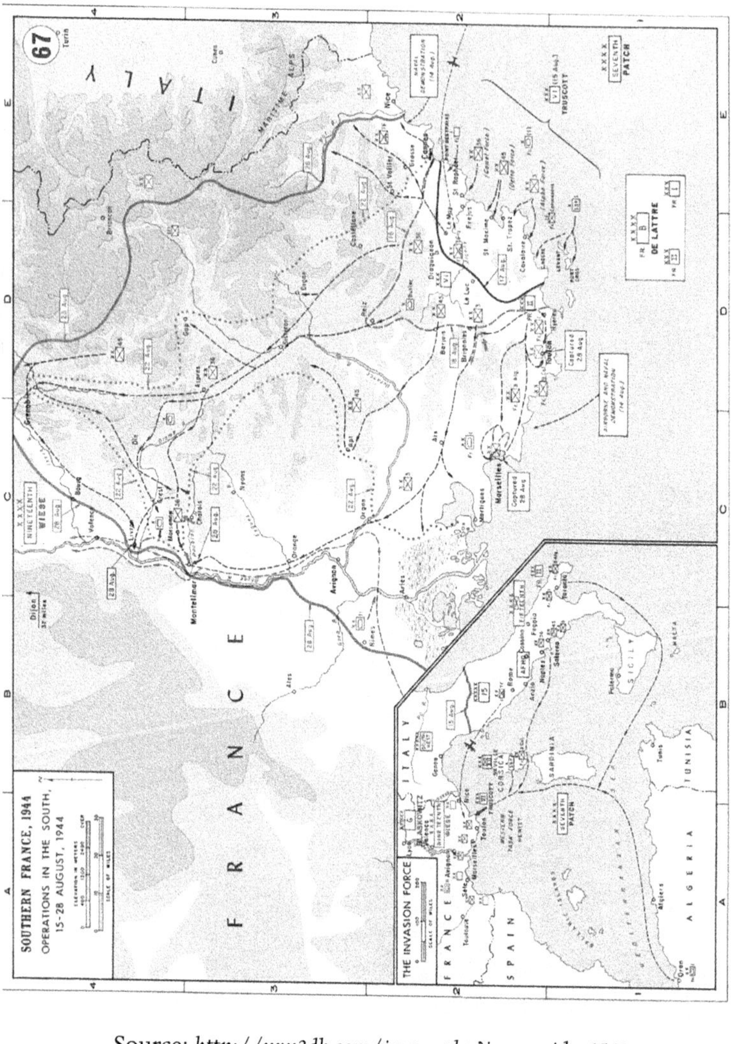

Source: *http://ww2db.com/image.php?image_id=6569*

Map 3

French Indochina
Source: *http://www.learnnc.org/lp/multimedia/13472*

APPENDIX C:
AARON BANK TIMELINE

1902	Nov 23—Born, New York City
1921-22	Attended Swedish Gymnastics Institute, New York City
1930-33	Lifeguard—Nassau, Bahamas and Biarritz, France
1938	Travel—Switzerland, England, Germany, Belgium
1942	Aug 19—Entered Army Service as a Private Dec 15—Entered Officer Candidate School
1943	Mar 15—Promoted to 2nd Lieutenant Oct 4—Promoted to 1st Lieutenant Aug 1—Joined OSS Nov 26—Left for European Theatre of Operations
1944	May 1—Promoted to Captain Jul 31 to Sept. 20—Team Packard Mission, France
1945	Jul 28—Arrived China Theatre of Operations Apr 22—Awarded Bronze Star Sep 11—Promotion to Major Dec 10—Assigned to Counter Intelligence Corps
1952	Jun 19—Appointed Inaugural Commander of 10th Special Forces Group
1958	Retired from U.S. Army
2004	Apr 1—Death

BIBLIOGRAPHY

Aaron Bank to Mrs. Beverly E. Lindsay. February 27, 1973. Alfred Paddock Private Collection.

Ambrose, Stephen E. *D-Day, June 6, 1944: The Climactic Battle of World War II*. New York: Simon & Schuster, 1994.

Ambrose, Stephen E. *Pegasus Bridge: June 6, 1944*. New York: Simon and Schuster, 1985.

Ballantine, Linda. Telephone interview by author. August 18, 2010.

Bank, Aaron. *From OSS to Green Berets: The Birth of Special Forces*. New York: Pocket Books, 1987.

Bank, Catherine. Telephone interview by author. August 18, 2010.

Beavan, Colin. *Operation Jedburgh: D-Day and America's First Shadow War*. New York: Viking, 2006.

Breuer, William B. *Daring Missions of World War II*. New York: J. Wiley, 2001.

Breuer, William B. *Operation Dragoon: The Allied Invasion of the South of France*. Novato, CA: Presidio Press, 1987

Brinkley, Douglas. *The Boys of Point Du Hoc: Ronald Reagan, D-Day, and the U.S. Army 2nd Ranger Battalion*. New York: Harper Collins Publisher, 2005.

Brownlee, Richard S. *Grey Ghosts of the Confederacy*. Baton Rouge: Louisiana State University Press, 1986.

Bull, Stephen. *Special Ops, 1939-1945: A Manual of Covert Warfare and Training*. Minneapolis, MN: Zenith Press, 2009.

Busch, Briton Cooper. *Bunker Hill to Bastogne: Elite Forces and*

American Society. Washington, D.C.: Potomac Books, 2006.

Carfano, James J. "Mobilizing Europe's Stateless: Americas Plan for a Cold War Army." *Journal of Cold War Studies* 1, no. 2 (Winter 1999): 1-38. *http://www.fas.harvard.edu/~hpcws/carafano. pdf* (accessed April 12, 2010).

Clancy, Tom, Carl Stiner, and Tony Koltz. *Shadow Warriors: Inside the Special Forces.* New York: G.P. Putnam's Sons, 2002.

Couch, Dick. *Chosen Soldier: The Making of a Special Forces Warrior.* New York: Crown Publishers, 2007.

Dommen, Arthur J., and George W. Dalley. "The OSS in Laos: The 1945 Raven Mission and American Policy." *Journal of Southeast Asian Studies* 22, no. 02 (September 1991): 327-46.

Dreux, William B. *No Bridges Blown.* Notre Dame, IN: University of Notre Dame Press, 1971.

Dwight D. Eisenhower to Collin Gubbins. May 31, 1945. *http:// www.spartacus.schoolnet.co.uk/SOEgubbins.htm* (accessed August 7, 2010).

Fitzgerald, Edward. E-mail interview by author. May 19, 2010.

Ford, Roger. *Steel from the Sky: The Jedburgh Raiders, France 1944.* London: Weidenfeld & Nicolson, 2004.

Funk, Arthur L. "Churchill, Eisenhower, and the French Resistance." *Military Affairs* 45, no. 1 (February 1981): 29-34.

Hans Waller, Barbara. Telephone interview by author. April 18, 2010.

Harrison, Gordon A. *Cross-Channel Attack.* Washington: Office of the Chief of Military History, Dept. of the Army, 1951.

Haswell, Jock. *The Intelligence and Deception of the D-Day Landings.* London: Batsford, 1979.

Healy, Joe. "SWCS Dedicates Bank Hall." *Special Warfare* 19, no. 1 (January/February 2006): 6-7. *http://www.soc.mil/swcs/swmag/ Archives/06_JanFeb.pdf* (accessed August 26, 2010).

Henry C. Lodge to General J. Lawton Collins. November 26,

1951. Massachusetts Historical Society, Boston.

Hogan, David W. *U.S. Army Special Operations in World War II*. Washington, D.C.: Center of Military History, Dept. of the Army, 1992.

Hogan, Jerry. "A Proud and Skillful Organization." *The Military View*, October 2008. *http://www.themilitaryview.com/?q=node/206* (accessed April 14, 2010).

Hogan, Jerry. "An Unusual Officer from Three Countries." *The Military View*, October 2008. *http://www.themilitaryview. com/?q=node/201* (accessed April 14, 2010).

Hogan, Jerry. Interview by author. April 14, 2010.

Irwin, Will. *The Jedburghs: The Secret History of the Allied Special Forces, France 1944*. New York: Public Affairs, 2005.

Joint Special Operations University and Office of Strategic Services (OSS) Society Symposium: Irregular Warfare and the OSS Model. Report. Hurlburt Field, FL: JSOU Press, 2010.

Kehoe, Robert R. "1944: An Allied Team With the French Resistance." *Studies of Intelligence*, Winter 98-99, 1-42. *https://www.cia.gov/library/center-for-the-study-of-intelligence/csi-publications/csi-studies/studies/winter98_99/art03.html* (accessed September 7, 2010).

Lewis, S. J. "Jedburgh Team Operations in Support of the 12th Army Group, August 1944." *CGSC—Command and General Staff College*. *http://www.cgsc.edu/carl/resources/csi/Lewis/Lewis. asp* (accessed August 17, 2010).

Lodge, Henry C. "Statement of Senator Henry Cabot Lodge." Speech, Address to First Group of Enlisted Aliens, Camp Kilmer, New Jersey, October 15, 1951. Massachusetts Historical Society, Boston.

Nathanson, E. M., and Aaron Bank. *Knight's Cross: A Novel*. New York: Carol Pub. Group, 1993.

Nutsch, Mark. Interview by author. April 17, 2010.

O'Donnell, Patrick K. *Operatives, Spies, and Saboteurs: The Unknown Story of the Men and Women of World War II's OSS*. New York: Free Press, 2004.

Paddock, Alfred H. *U.S. Army Special Warfare: Its Origins*. Lawrence, KS: University Press of Kansas, 2002.

R. W. Volckmann to Mrs. Beverly E. Lindsay. March 21, 1969. Alfred Paddock Private Collection.

Records of the OSS. RG 226. Entry 103. Box 3. Folder 78, Team Packard. National Archives. College Park, MD.

Records of the OSS. RG 226. Entry 154. Box 174. Field Station Files: Kunming. National Archives. College Park, MD.

Records of the OSS. RG 226. Entry 154. Box 199. Field Station Files: Kunming. National Archives. College Park, MD.

Records of the OSS. RG 226. Entry 154. Box 202. Field Station Files: Kunming. National Archives. College Park, MD.

Records of the OSS. RG 226. Entry 174. Box 154. Field Station Files: Kunming. National Archives. College Park, MD.

Records of the OSS. RG 226. Entry 224. Box 34. Personal File: Bank, Aaron. National Archives. College Park, MD.

Records of the OSS. RG 226. Entry 92a. Box 87. Field Station Files: Kunming. National Archives. College Park, MD.

Records of the OSS. RG 226. Folder 3371. Box 198. Field Station Files: Kunming. National Archives. College Park, MD.

"Report of Jedburgh Team Alexander." Digital image. Operation Jedburgh. *http://www.operationjedburgh* (accessed August 9, 2010).

"Report of Jedburgh Team Bruce." Digital image. Operation Jedburgh. *http://www.operationjedburgh* (accessed August 9, 2010).

"Saving Lives, Destroying Them—— All in His Colorful Past." Interview by Gordon Grant. *Los Angeles Times* (Los Angeles), August 11, 1968.

Schoenbrun, David. *Soldiers of the Night: The Story of the French*

Resistance. New York: Dutton, 1980.

Sides, Hampton. *Ghost Soldiers: The Forgotten Epic Story of World War II's Most Dramatic Mission.* New York: Doubleday, 2001.

Simons, Laval. Interview by author. April 17, 2010.

Simpson, Charles M. *Inside the Green Berets: The First Thirty Years, a History of the U.S. Army Special Forces.* Novato, CA: Presidio Press, 1983.

Singlaub, John K. *Hazardous Duty: An American Soldier in the Twentieth Century.* New York: Summit Books, 1991.

Smith, R. Harris. *OSS: The Secret History of America's First Central Intelligence Agency.* Berkeley: University of California Press, 1972.

Southworth, Samuel A. *U.S. Special Warfare: The Elite Combat Skills of America's Modern Armed Forces.* Cambridge, MA: Da Capo Press, 2004.

Speidel, Hans. *Invasion 1944; Rommel and the Normandy Campaign.* Westport, CT: Greenwood Press, 1971.

Sutherland, Ian. "The OSS Operational Groups: Origin of Army Special Forces." *Special Warfare* PB 80-02-2 (June 2002): 2-13. *http://www.soc.mil/swcs/swmag/Archives/02jun.PDF* (accessed October 13, 2010).

Thomas, David. "The Importance of Commando Operation in Modern Warfare 1939-82." *Journal of Contemporary History* 18, no. 4 (October 1983): 689-717.

Tovar, Hugh. E-mail interview by author. November 8, 2010.

"Transcipt of Charles De Gaulle 6-18-40 Speech." The Lehrman Institute Public Policy Programs Lehrman Institute Research. *http://lehrmaninstitute.org/history/index.html* (accessed October 13, 2010).

United States. Dept. of the Army. *FM 31-21 Guerilla Warfare and Special Forces Operations, 1961.* Washington: U.S. Government Printing Office, 1961.

U.S. Congress. House. Extension of Remarks. *Tribute to Col.*

Aaron Bank. By Ron Packard. 104th Cong., 1st sess. H. Res. E 389. *http://www.gpo.gov/fdsys/pkg/CREC-1995-02-21/pdf/CREC-1995-02-21-extensions.pdf* (accessed August 9, 2010).

U.S. Congress. Senate. 1950. *Public Law 597*, S.S. 2269, 81st Cong., 2d sess. Accessed 12 Jan 2011; available from GPO Access.

Volckmann, R. W. *We Remained: Three Years behind the Enemy Lines in the Philippines.* New York: W.W. Norton, 1954.

Waller, Douglas C. *The Commandos: The Inside Story of America's Secret Soldiers.* New York: Simon & Schuster, 1994.